Statistical Modelling
with
Quantile Functions

Statistical Modelling
with
Quantile Functions

CRC Press
Taylor & Francis Group
Boca Raton London New York

CRC Press is an imprint of the
Taylor & Francis Group, an **informa** business

CRC Press
Taylor & Francis Group
6000 Broken Sound Parkway NW, Suite 300
Boca Raton, FL 33487-2742

First issued in paperback 2019

ISBN-13: 978-1-58488-174-2 (hbk)
ISBN-13: 978-0-367-39868-2 (pbk)

Library of Congress Cataloging-in-Publication Data

Gilchrist, Warren, 1932-
    Statistical modelling with quantile functions / Warren G. Gilchrist.
        p. cm.
    Includes bibliographical references and index.
    ISBN 1-58488-174-7 (alk. paper)
    1. Distribution (Probability theory) 2. Sampling (Statistics) I. Title.

QA276.7 .G55 2000
519.2—dc21

00-023728
CIP

Library of Congress Card Number 00-023728

Visit the Taylor & Francis Web site at
http://www.taylorandfrancis.com

and the CRC Press Web site at
http://www.crcpress.com

Contents

x

List of Figures

List of Tables

Preface

In the early 1980s while working on a book on statistical modelling (Gilchrist (1984)), I came across a distribution called the generalized lambda distribution. I duly mentioned it in passing, (p. 45), observing that it was defined in a different manner than the other distributions dealt with in the book. Later on, while working with students on placement with Glaxo-Chem and in other work with Glaxo Wellcome, the opportunity arose to become familiar with the practical use of this highly flexible distribution. It slowly dawned on me, from this work and beginning to read the scattered literature of relevance, that the evident practical value of the methods used was not just a matter of one distribution but of a particular way of looking at statistical modelling. As a result, the ideas in this book were slowly put together. The book is not intended as a research monograph, but rather as a straightforward introductory text for practitioners of statistical modelling. However, the approach adopted is different than the standard approach, as was taken in the previous book, and leads to different definitions of even basic quantities like the mean. Thus, in many ways we are forced to go back to basics and look at statistical modelling from the beginning. We look, however, from a different perspective, which is not intended as a replacement to the classical approaches to modelling but as a useful supplement, and indeed it will be seen that there are many overlaps.

I would like to thank the following individuals: staff from Shef-field Hallam University, especially Dr Steve Salter, Richard Gibson, and Penny Veitch; Professor Emanuel Parzen for his helpful comments and particularly for the idea of starting with an overview chapter; staff from the Glaxo Wellcome Group, particularly Dr Max Porter and Miss Gillian Amphlett; Dr Alan Hutson and James Meginniss for their many helpful comments on the penultimate draft text (although any errors in the final text are my responsibility); and my wife for her support and her patience with a husband who is supposed to be retired.

Warren Gilchrist
January 2000
w.g.gilchrist@shu.ac.uk

CHAPTER 1

An Overview

1.1 Introduction

In the balance of this book we will look systematically at the many issues associated with the steps of the statistical modelling process, using an approach based on what will be termed *quantile methods.* However, there is the danger that, if we do just go through these methods step by step, the reader will not see the wood for the trees. The aim of this chapter is thus to provide an outline of the wood. This will be done by bringing together the simplest of the ideas and approaches of quantile modelling to form both an initial picture and a basis for later development.

Anyone who has studied statistics, even at the most basic level, will have already met quantile methods, although may not be aware of it. A quantile is simply the value that corresponds to a specified proportion of a sample or population. Thus the median of a sample of data is the quantile corresponding to a proportion 0.5 of the ordered data. For a theoretical population it corresponds to the quantile with probability of 0.5. If $Q(p)$ is the function, called the **quantile function,** that gives the quantile values for all probabilities p, $0 \le p \le 1$, then the median is $Q(0.5)$. Similarly, we have the quartiles $Q(1/4)$ and $Q(3/4)$. Most users of statistics will have utilized tables of the normal distribution to look up, for example, values such as 1.96 as the value that has a probability of 0.975 of not being exceeded. Thus if $N(p)$ is the quantile function for the standard normal distribution then 1.96 is $N(0.975)$, so the normal tables used are just the tables of the quantile function for the standard normal distribution.

What is the link then to modelling? The fact that in the last discussion a standard normal quantile function was used implied that there was some prior modelling to justify the use of the normal distributional model rather than any one of the multitudes of alternatives. The fact that a standard normal distribution was used required some matching of the data being studied to the normal distribution model. Almost all practical statistics imply some form of modelling.

We will study these ideas again more carefully later. But we still have not said why an entire book needs to be devoted to linking the ideas of quantile functions to the process of statistical modelling. There are two prime reasons:

1. Statistical modelling is a tool used essentially as part of problem solving. The purpose of having a statistical model is almost always to assist in some problem-solving activity. The model assists in carrying out a prediction; it is used in some selection process or it helps in answering some "what if ...?" question. Modelling is part of the problem solving process. A well-recognised feature of problem solving is that problems are often solved by seeking alternative ways of looking at the problem. For example, where data is a series of observations in time, the classical models use a time origin, $t = 0$, that is a fixed point in time. If, however, $t = 0$ is a moving time, e.g., it is always $t =$ now, then a new range of models become available for studying the problems of time series. It will be seen repeatedly through the pages of this book that expressing statistical ideas in terms of quantile functions gives both a new perspective and sometimes a simpler and clearer perspective. Thus an approach based on quantile functions provides an additional perspective for the problem solver in the use of statistical models.

2. Many statistical models consist of deterministic and stochastic (chance, random) elements. The classical approach to statistical modelling is such that the deterministic element is often built up by adding together, or sometimes multiplying, simple components, as with a construction kit like Lego™. The stochastic element, however, is usually chosen from a library of distributional models that has been built up over the last few centuries and is described in such books as Johnson, Kotz and Balakrishnan (1994 and 1995) and Evans, Hastings and Peacock (1993). We will see that if the stochastic element is modelled using quantile functions, then both elements in the model can be developed with a common construction kit approach. In the same way that deterministic modelling is a construction process so, too, using quantile functions is distributional modelling. Together there is a unified approach to model construction.

For the rest of this text we will amplify and explore these two ideas. For now, however, let us start at the beginning.

1.2   The data and the model

Statistics has to start with data, a set of numbers collected in some experiment, survey or observational study. Sometimes a simple study of the data may seem to tell us what we want to know. We can draw plots of the data and work out average values and get a feel for the situation being studied. However, one feature that is almost universally present in data is its variability. In addition to the factors in the situation of which we are aware, there will be many other chance influences jostling data values away from the perfect information we would like to have. We are forced to recognise that if we repeated the exercise we would not get a repeat of the same set of data. We thus have to speak of the data obtained as being a **sample** from a **population** of possible values. The language reflects the early use of a sample of people being measured as representative of the whole population of a country. It is hoped, for example, that a large random sample will clearly and accurately show the features of the population of interest. We can describe the sample using graphs and summary numbers, such as the average. To describe the population we need the concept of the **model**, which is usually a mathematical description of the features of interest. Over the centuries, vast ranges of mathematically based models have been developed to cover situations in both sciences and social sciences. These models often consist of two components. First, there is a **deterministic** element which describes behaviour or relationship with no allowance for chance or variability. Second, there is the **random** element that allows for the influence of the uncertainty inherent in almost all situations. The focus of this book is on the forms that the random element may take, although in Chapter 12 we look at the construction of models that involve the two elements.

1.3   Sample properties

Before we can sensibly model a set of data we need to have a clear perception of it, otherwise we will find ourselves imposing our views of how it ought to behave, rather than finding out how it does behave. The best way to develop this perception and feel for the data is graphically, with the support of some numerical summaries of the properties seen on the graphs. Failure to link the graphs and summaries can result in the use of summaries that are meaningless or misleading for the data being analyzed. Let us look at a set of data to illustrate a number of different approaches.

Example 1.1: Table 1.1 gives the layout for a calculation based on the maximum flow of flood water on a river for 20 periods of 4 years, in million cu. ft. per sec. The data are given and studied in Dumonceaux and Antle (1973). There are no strong features over time so the distributional features of the whole set of data will be used for illustration, although we will return to this issue later. The original data has been sorted by magnitude and placed in the column headed x. This is a fundamental step in the types of analysis that will be developed in this text. It is the 'shape' of the ordered data that describes its structure. With each ordered value, x, we will associate a probability p, indicating that the x lies a proportion, p, of the way through the data. At first guess we would associate the rth ordered observation, denoted by $x_{(r)}$, with p_r

p	x			$n = 20$			
0.025	0.265	Dx	Dp	Mid-x	Mid-p	Dx/Dp	Dp/Dx
0.075	0.269	0.004	0.05	0.267	0.05	0.08	12.50
0.125	0.297	0.028	0.05	0.283	0.1	0.56	1.79
0.175	0.315	0.018	0.05	0.306	0.15	0.36	2.78
0.225	0.3225	0.008	0.05	0.319	0.2	0.15	6.67
0.275	0.338	0.016	0.05	0.330	0.25	0.31	3.23
0.325	0.379	0.041	0.05	0.359	0.3	0.82	1.22
0.375	0.380	0.001	0.05	0.380	0.35	0.02	50.00
0.425	0.392	0.012	0.05	0.386	0.4	0.24	4.17
0.05	0.402	0.010	0.05	0.397	0.45	0.20	5.00
0.525	0.412	0.010	0.05	0.407	0.5	0.20	5.00
0.575	0.416	0.004	0.05	0.414	0.55	0.08	12.5
0.625	0.418	0.002	0.05	0.417	0.6	0.04	25.00
0.675	0.423	0.005	0.05	0.421	0.65	0.10	10.00
0.725	0.449	0.026	0.05	0.436	0.7	0.52	1.92
0.775	0.484	0.035	0.05	0.467	0.75	0.70	1.43
0.825	0.494	0.010	0.05	0.489	0.8	0.20	5.00
0.875	0.613	0.119	0.05	0.554	0.85	2.38	0.42
0.925	0.654	0.041	0.05	0.634	0.9	0.82	1.22
0.975	0.74	0.086	0.05	0.697	0.95	1.72	0.58

Table 1.1. Layout for flood data plots

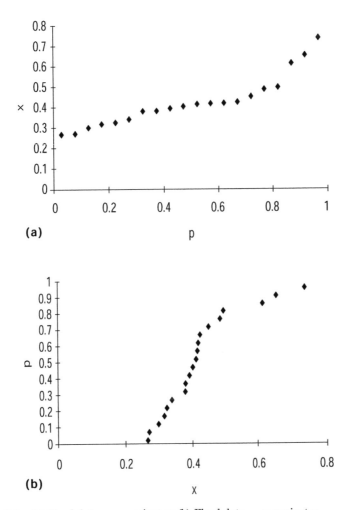

Figure 1.1. (a) Flood data — x against p; (b) Flood data — p against x

$= r/n$, $n = 20$, $r = 1,...,20$. However, the range of values we would expect for p is $(0, 1)$. The value of r/n, however, goes from $1/20$ to 1, i.e., it is not symmetrical. Hence, to get the p values symmetrically placed in the interval $(0, 1)$, we use the formula $p_r = (r - 0.5)/n$. This formula corresponds to breaking the interval $(0, 1)$ into 20 equal sections and using the midpoint of each. Thus we have pairs of values describing the data as $(x_{(r)}, p_r)$. The value of x for any p is referred to as the **sample p-quantile**. There are two natural plots of such data. First, of $x_{(r)}$ against p_r and second, of p_r against $x_{(r)}$, which are just the same plot with the

axes interchanged. However, a look at these plots in Figure 1.1(a) and
(b) shows that different features stand out most clearly in different plots.
For x against p, there is a steady increase in x that becomes a steeper
increase towards the higher values of p. For the plot of p against x, the
change in the slope of p with x looks more dramatic, as does the break
in the data. The break is due to two of the ordered observations being
somewhat further apart than those around them. This is almost certainly
a chance feature and often occurs in such plots. The difference in slope
between high values and low values of x looks to be a much more natural
feature of the data.

The discussion of slopes suggests that the slopes themselves be plotted.
Thus if Dx is the difference between two successive values of the ordered
x, called spacings, and Dp is the difference in their p values, then we
can derive a set of 19 pairs (Dx,Dp). The calculation is shown in
Table 1.1 and the consequent slopes Dx/Dp and Dp/Dx calculated. Figure
1.2(a) shows Dp/Dx plotted against the mid-values of the two xs used
for each Dx. Figure 1.2(b) shows the values of Dx/Dp plotted against
the mid-values of the two ps used for each Dp. In Figure 1.2(a), the
points are linked to form a polygon. Linking successive points is a
procedure that sometimes helps, and sometimes hinders, seeing the
data. The high peak is due to two x values being particularly close
together. The plot shows a great deal of random behaviour, but also
some structure, with the higher values being around 0.4 and a strag-
gling tail to the right. The plot of Dx/Dp against p shows low values
over the central third of the probability and higher values at the
extremes, particularly to the right. One final variant used in plotting
these quantities is to plot Dp/Dx, not against x, but against p. This plot
shows how the rate of change in the probability is modified according
to how far, proportionately, the observations are through the data, as
Figure 1.3 illustrates.

The problem of the randomness exhibited by Dp/Dx is partially dealt
with here by a process of smoothing the data. We deal with the detail
of this later. It clearly helps the picturing of the situation, at the cost of
some distortion. This group of five plots illustrates the points that (a)
in any set of data, particularly when there are as few as 20 observations,
there will be many results of sheer randomness in the data; (b) different
plots will show most clearly different features of the data; (c) in spite of
the randomness, there are features, structures, in the data that the user
can seek to study and describe.

Before leaving the data presented in Table 1.1, it should be noted that
during the course of this book we will often show the layouts for calculations, sometimes just a section of a bigger table for illustration. The

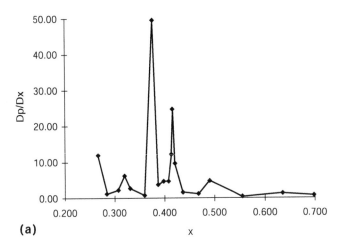

(a)

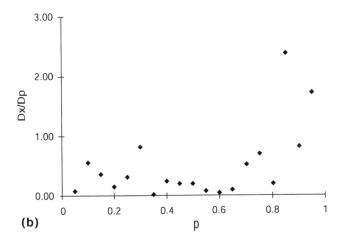

(b)

Figure 1.2. (a) Flood data — *Dp/Dx* against mid-*x*; (b) Flood data — *Dx/Dp* against mid-*p*

approach throughout is to present statistical calculations in column form. This approach is regarded as essential for building a feel for both the data and the models being used. It also facilitates the plotting of graphs, another central emphasis in our studies.

In addition to graphical studies of data it is useful to have some summary values that give a numerical indication of the shape of the data. The simplest of these is the **sample median**, *m*, which is the mid-value in the data. With the even number of observations in Table 1.1 the median

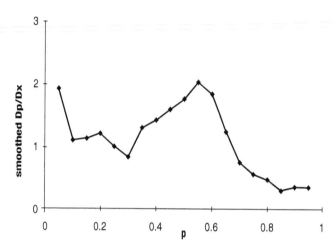

Figure 1.3. Flood data — smoothed *Dp/Dx* against mid-*p*

is taken to be midway between the tenth and eleventh observations, so $m = (0.402 + 0.412)/2 = 0.407$. The values one quarter and three quarters of the way through the data are called the **sample lower quartile** and **upper quartile**, respectively, *lq* and *uq*. For Table 1.1 these will again lie between data points. We use the halfway point for now, but discuss the matter further later. Thus $lq = (0.3225 + 0.338)/2$, which is 0.330. Similarly, $uq = 0.4665$.

Two measures of the shape of the data can be derived from the quartiles. The first is the **interquartile range**, *iqr*, which is simply the difference between the two quartiles, $iqr = uq - lq$. For Table 1.1, this is 0.136. The *iqr* gives a simple measure of the spread of the data. Parzen (1997) suggests the **quartile deviation**, defined as *2iqr*, as a more suitable quantity for some purposes. The **range**, *w*, of the data is the total spread from smallest to largest observation, which in our case is 0.475. For a symmetrically shaped distribution, the deviations of the quartiles from the median will be approximately the same. Thus the difference between these deviations is a measure of the non-symmetry, the **skewness**, of the data. The second quartile-based measure defines the **quartile difference** by

$$qd = (uq - m) - (m - lq) = lq + uq - 2m$$

For the data of Table 1.1, this is –0.018. A standard measure of skewness is given by **Galton's skewness coefficient**, *g*, which divides the quartile difference by the inter-quartile range to get a measure that does not depend on the measurement scale of the data. Thus $g = qd/iqr$. For the data here, this is −0.13, which is a small skewness to the left. A look at Figure 1.2(a) reveals the care needed when handling single measures such as g. The value of g depends on the values of the observations at or next to three specific points. From the figure it is seen that to the right of these there are three observations extending well to the right. There is thus some right skewness that this measure misses for this data. This fact emphasizes the need to visually study data as well as develop summaries. It also points to the need for a wider range of summaries, which will be addressed later.

The above example introduces a number of ways of measuring and plotting some of the features that are often important in understanding a set of data. Just having such information alone may have practical use. However, we are often seeking to achieve some model of the situation, some theoretical model that has a structure similar to that shown by the data. We want to link the sample features to corresponding features in the population from which our sample comes. We therefore need to turn to ways of describing a population.

1.4   Modelling the population

In this section we will concentrate on the random component of statistical models and in particular on the models for the random variability involved. Figures 1.1(a) and (b) and 1.2(a) and (b) illustrated four ways that can be used to show the structure of a sample from a distribution. In population terms these represent four different ways in which we may define the model for the random variation. The first two of these will be found in almost all statistics texts, the second two are rarely mentioned. We will describe all four and also introduce some specific illustrative models for future use.

The cumulative distribution function

The Cumulative Distribution Function, CDF, denoted by $F(x)$, is defined as the probability of a variable X being less than or equal to some given value, x, using the convention of capitals for a random variable generally and lower case for specific values. Thus

$$F(x) = \text{Probability (the random variable } X \leq x).$$

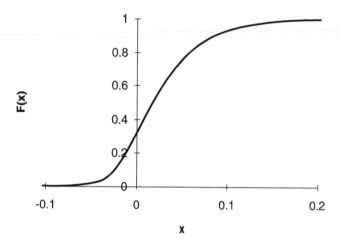

Figure 1.4. A cumulative distribution function, $F(x)$

Figure 1.4 shows a typical form for such a function. The function must start at the value of zero probability at the left-hand end of the axis and build up to the probability of one (certainty) to the right. It must also clearly be a non-decreasing function. The plot corresponds to the sample plot of p against x. We refer to the way the probabilities associated with a variable spread over the possible values as the **distribution**, however they are formally defined. The CDF thus gives one way of defining a distribution.

> **Example 1.2:** Many pocket calculators and most, if not all, mathematical or statistical software provide a function usually called RAND. This function produces a number in the interval (0, 1). Each time it is used it produces a different number, all numbers being equally likely. These are **random numbers**. If U is such a number, it is said to have a continuous **uniform distribution**. For such a distribution, the probability of a number being less than 0.5 will be 0.5, etc. Hence in formal terms and using the convention of u for a uniform variable:
>
> $$F(u) = u, \ 0 \leq u \leq 1.$$
>
> To this we would technically need to add that
>
> $$F(u) = 0, \ u < 0, \ F(u) = 1, \ u > 1,$$
>
> but for practical purposes we keep to values in the range of the distribution given by $0 \leq u \leq 1$.

Example 1.3: A distribution that frequently occurs where time is the variable, such as the time interval between arrivals of customers at a service point, is called the **exponential distribution**. The probability of arrival before time x increases most rapidly at first and then gradually flattens towards its ultimate value of one. The mathematical form that describes this curve is:

$$F(x) = 1 - e^{-\gamma x}, \ 0 \leq x \leq \infty, \ \gamma \geq 0.$$

The parameter γ represents in our illustration the rate of arrival of customers.

The probability density function

The Probability Density Function, PDF, denoted by $f(x)$, provides a second means of defining a distribution. Formally it is defined by the relation:

$$f(x)dx = \text{Probability } (x \leq \text{the random variable } X \leq x + dx),$$

where dx is an infinitesimally small range of x. The area under the curve of $f(x)$, which is the total probability of having any observed value, must be one. To see the relation between the CDF and PDF note that we can write, in simple calculus terminology,

$$f(x)dx = \text{Probability (observation in } dx) = F(x + dx) - F(x) = dF(x).$$

Thus the PDF, $f(x) = dF/dx$, is the derivative of the CDF. Figure 1.5 illustrates the form of a PDF. Note that this population function corresponds to the sample plot of Dp/Dx against x.

Example 1.4: For the uniform distribution $F(x) = x$, so $f(x) = 1$. More formally we write

$$f(x) = 1. \ 0 \leq x \leq 1.$$
$$= 0, \text{ otherwise.}$$

Example 1.5: For the exponential distribution, the derivative of the CDF is $\gamma e^{-\gamma x}$, hence the PDF is

$$f(x) = \gamma e^{-\gamma x}, \ 0 \leq x \leq \infty$$
$$= 0, \ -\infty \leq x < 0.$$

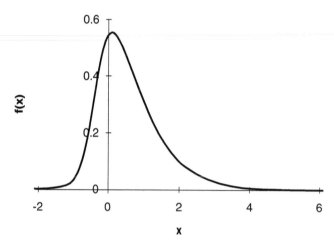

Figure 1.5. A probability density function, $f(x)$

As mathematics has developed it has ensured that all well-tabulated and used functions have derivatives that are well tabulated and used. Hence if a distribution has a CDF that can be explicitly expressed in terms of common functions, it will also have a PDF that can be so expressed. However, the converse is not the case:

Example 1.6: One of the most commonly occurring distributions is the **normal distribution,** which is defined by

$$f(x) = [1/\sqrt{(2\pi)}\sigma] \exp[-(x - \mu)^2/2\sigma^2], \ \infty \le x \le \infty.$$

the parameters μ and σ control the positioning and spread of the distribution. Unfortunately, this PDF does not have a CDF that can be written in simple explicit terms. For the case where $\mu = 0$ and $\sigma = 1$, the distribution is called the **standard normal distribution**. The CDF has to be evaluated numerically, but is available in statistical tables, statistical software, and spreadsheets.

The quantile function

The Quantile Function, *QF*, denoted by $Q(p)$, provides a third way of defining a distribution. Formally, we have

$$x_p = \text{the value of } x \text{ for which Probability } (X \le x_p) = p.$$

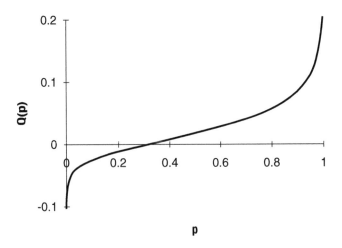

Figure 1.6. A quantile function, $Q(p)$

The value x_p is called the **p-quantile** of the population. The function $x_p = Q(p)$ expresses the p-quantile as a function of p and is called the **quantile function**. The first paper to systematically develop quantile functions was by Parzen (1979). The definitions of the QF and the CDF can be written for any pair of values (x, p) as $x = Q(p)$ and $p = F(x)$. These functions are thus simple inverses of each other, provided that they are both continuous increasing functions. Thus we can also write $Q(p) = F^{-1}(p)$ and $F(x) = Q^{-1}(x)$. For sample data the plot of $Q(p)$ is the plot of x against p. Figure 1.6 shows the quantile function corresponding to the CDF of Figure 1.4.

Example 1.7: Putting $F(x) = p$ in the uniform distribution, so $p = x$, gives as the inverse

$$Q(p) = p, \ 0 \le p \le 1.$$

Notice that $Q(p)$ is only defined for the interval $0 \le p \le 1$, since probabilities do not exist outside this range, even if the mathematical function does. This condition will be implicit in all statements involving p for the rest of this book.

Example 1.8: If we take the CDF of the exponential distribution, $F(x) = 1 - e^{-\gamma x}$, write $F(x) = p$ and solve the equation for x in terms of p, we get

$$Q(p) = -\eta \ln(1 - p), \text{ where } \eta = 1/\gamma.$$

It will be seen that the parameter η controls the spread of the distribution. It is a **scale parameter**.

Example 1.9: Consider the distribution defined by its QF:

$$Q(p) = 0.2 \ln(p) - 0.8 \ln(1 - p).$$

This quantile function is an example of a distribution called the **skew logistic**. The shape of the quantile function is shown in Figure 1.6. We will discuss it later. For now it is sufficient to note that we cannot invert this quantile function to get p in terms of x, so that no explicit CDF, or consequently PDF, exists. The PDF can, however, be drawn and Figure 1.5 shows its shape.

Example 1.10: The commonly used normal distribution does not have an explicit QF and numerical methods are required to evaluate it. Nonetheless, it is widely used in statistical tests and forms the basis of many statistical tables. Statistical software and spreadsheets will give the values of the standard normal QF, $N(p)$. To obtain the quantiles of a normal distribution with parameters μ and σ, we use the quantile function $Q(p) = \mu + \sigma N(p)$. The standard normal is symmetric about the value zero, so $Q(p)$ will be symmetric about the parameter μ, the mean, which controls position, and σ, the standard deviation, which controls the scale.

The quantile density function

In the same way that the CDF can be differentiated to give the PDF, we can use the derivative of the QF to obtain a function of considerable value in describing distributions. This derivative is the Quantile Density Function, QDF, defined by

$$q(p) = dQ(p)/dp.$$

As $Q(p)$ is a non-decreasing function it follows that its slope, $q(p)$, is non-negative, at least for $0 \le p \le 1$. We have seen the corresponding form of $q(p)$ for sample data in the plot of Dx/Dp against p.

Example 1.11: For three of the distributions previously mentioned that have explicit $Q(p)$, we obtain

Uniform: $q(p) = 1$.

Exponential: $q(p) = 1/(1 - p)$.

Skew Logistic: $q(p) = 0.2/p + 0.8/(1 - p)$.

We now have four different, but related, ways of defining a distribution. Classical statistics is based on using the first two and almost all distributions are defined in terms of their CDF or PDF. As we have seen, there are some distributions like the exponential that have explicit mathematical forms for each of the CDF, PDF, QF and QDF. However, there are others that only have a simple mathematical form in one or two of the four; for example, the skew logistic that we have illustrated has only explicit QF and QDF. If we wanted values of the CDF or PDF for these distributions, use would have to be made of numerical techniques to obtain specific approximate values of p for a given x. In this book we concentrate primarily on the use of the quantile function for the formulation of distributions. As we have noted, the quantile function is rarely mentioned explicitly in statistical textbooks. It is often used, however, since a little study will show that most statistical tables are, in fact, tables of the quantile functions for various distributions. As a shorthand terminology we refer to distributions of any type that we approach from the quantile point of view as **quantile distributions**. In the next section we will explore the justification for approaching statistical modelling from a basis of quantile distributions.

1.5   A modelling kit for distributions

When in the usual process of statistical modelling the deterministic part of a model is considered, the model is developed using what can best be described as a modelling kit approach. For example, suppose one has a simple set of monthly profit figures, x_t, over several years of data. A plot of the data might suggest that although there is a general mean level over the period, there are also an upward trend, an annual seasonal variation, and some random, irregular variation. To model these features we would introduce terms and components: L for the average level, T for the trend (the slope), S for the seasonal variation, and I for the random variation. Without going into the definitions and details it is intuitively reasonable that these could then be combined to form a reasonable model. This might be

$$x = L + T + S + I,$$

or if the seasonal effect creates a percentage change in the underlying mean and trend levels, possibly

$$x = (L + T)S + I.$$

Thus the model consists of a set of simple components that can be added or multiplied to create an appropriate model for the data. The components and the joining rules of addition and multiplication can be regarded as a modelling kit analogous to Lego™ or Meccano™. When we look specifically at the random component I, then the chances are some common distribution, such as the Normal, will be used.

The modelling of random variability almost always takes the form of the selection of a distribution from a large library of distributional models. This library has been built over the last two hundred years or so. A transformation of the data may be used if the library model does not work too well, but unfortunately this transforms the deterministic part of the model as well, which may create its own difficulties. The reason for the difference in approach to the deterministic and the random aspects of the model is that, unlike the components of the deterministic part of a model, we cannot add or multiply components for the random part of a model and get a developed model, at least not when they are defined in the common way by CDF or PDF. Addition of CDF or PDF leads to mixtures of distributions rather than new distributions. Thus there is no model building kit available for the random component parallel to that for the deterministic component. Modellers are therefore obliged to use ready-made distributional models from the library. The basis for the approach to statistical modelling based on quantile function models, developed in this text, is that *quantile functions can be added and, in the right circumstances, multiplied to obtain new quantile functions. Quantile density functions can also be added together to derive new quantile density functions.*

These properties enable the modeller to seek an appropriate model for a distribution by combining component models to create new models. We will also see later that, as with deterministic model components, we can develop quantile functions by transformations of various types. Thus there does exist a modelling kit for the distributional element in models and it is based on quantile functions. The statistical modeller is thus in the same situation in relation to both deterministic and random components for the model. On the basis of the data, the task is to choose appropriate deterministic and random components and find appropriate ways of combining them so that the fitted model will show the same basic properties as are observed in the data. The modeller is thus not limited to the library of classical statistical models but can tailor the models to the needs of the specific application. To prepare the ground for this we first need to look more closely at the quantile function and its elementary properties.

1.6   Modelling with quantile functions

As a means of introducing the basic modelling properties of the quantile function and quantile density function we will work through a sequence of examples.

Example 1.12: Let us start by returning to the exponential distribution. The mathematical form of the PDF is almost always used as the basic definition in texts, thus

$$f(x) = \gamma e^{-\gamma x}, \, x \geq 0.$$

As we have seen, the CDF and QF are $F(x) = 1 - e^{-\gamma x}$ and $Q(p) = -\eta \ln(1 - p)$, where $\eta = 1/\gamma$. The range of values that x can take, the **distributional range or limits**, DR, is found by putting the probability range of 0 to 1 in $Q(p)$. Thus $Q(0) = 0$ and $Q(1) = \infty$. To get the middle value of the distribution, the **population median**, M, put $p = 0.5$, which gives $M = -\eta \ln(0.5) = \eta \ln(2)$. The sample median, m, has already been introduced. As a matter of convention, upper case letters will be used for population measures and lower case letters for the corresponding sample measures.

Example 1.13: Consider the quantile function $Q(p) = \lambda - \eta \ln(1 - p)$. If $\lambda = 0$ and $\eta = 1$, then we have $S(p) = -\ln(1 - p)$, which is the unit exponential distribution and for which $M = \ln(2)$. It is evident that multiplying by η stretches the scale of the distribution. Thus η is a **scale parameter**. If the parameter λ is now added, the left-hand end of the distributional range becomes $Q(0) = \lambda$ and the median becomes $M = \lambda + \eta \ln(2)$. Thus the whole distribution moves its position an amount λ to the right. The parameter λ is hence called a **position parameter**. In this particular example the position parameter is also a **threshold parameter** since it defines the lower end of the distributional range.

The last example illustrates the general ability of **QF** approaches to change position and scale in a simple way. It is, therefore, convenient to first look at distributions, without reference to position or scale, by using the simplest, most basic formula, such as $-\ln(1-p)$ for the exponential. We will call these simplest formulae the **basic forms**, and denote their quantile functions by $S(p)$. $S(p)$ may contain parameters that control shape. The basic form can then be generalized to include position and scale parameters by using the transformation

$$Q(p) = \lambda + \eta\, S(p).$$

In future discussions we will use $S(p)$ where it is appropriate to highlight that we are using a basic form.

Example 1.14: Suppose a set of data is obtained from the unit exponential distribution and then has a minus sign put in front of each observation. The consequent distribution will be called the **reversed** or **reflected exponential**, shown in Figure 1.7. The quantile function for the ordinary unit exponential is, as has been seen, $S(p) = -\ln(1-p)$. Looking at the figure it is clear that the p-quantile for the reflected exponential is minus the $(1-p)$-quantile for the ordinary exponential. Thus the quantile function for the reflected exponential is $S(p) = \ln(p)$. For $p = 0$, this is $-\infty$ and for $p = 1$, it is zero as required. Figure 1.8(a) shows the quantile functions of the exponential and its reflected counterpart.

Example 1.15: Suppose the quantile functions of the exponential and the reflected exponential are added together giving

$$S(p) = -\ln(1-p) + \ln(p) = \ln[p/(1-p)].$$

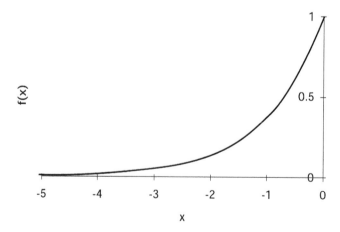

Figure 1.7. PDF of the reflected exponential

Figure 1.8 (b) shows a plot of this sum. It will be seen that the centre
line, obtained by the addition, is symmetrical about the value $x = 0$,
which occurs at $p = 0.5$ and therefore is the median. It is also evident
that in the tails the shape is that of the corresponding exponential or
reflected exponential distribution. This model is called the **logistic distribution**. Figure 1.9 shows the addition operation for the quantile
density functions of the exponential and reflected exponential. As the
derivative of a sum is the sum of the derivatives, the addition of quantile

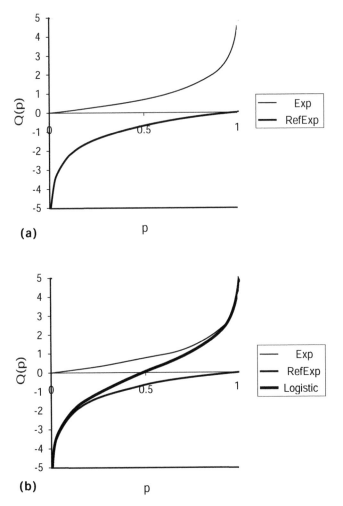

Figure 1.8. (a) Quantile functions of the exponential and reflected exponential.
(b) Addition of exponential and reflected exponential quantile functions

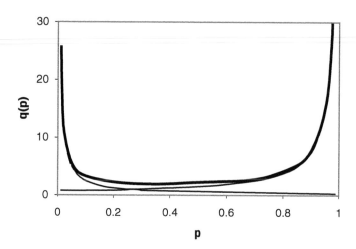

Figure 1.9. Addition of quantile density functions

functions must lead to the addition of quantile density functions. Figure 1.10 shows the PDF of the logistic distribution.

Example 1.16: Return briefly to the uniform distribution, for which the quantile function is $S(p) = p$. Obviously the distributional range for this distribution is (0,1) and the median is 0.5.

Example 1.17: Suppose a set of data has exponential-looking tails but is much flatter at the peak than the logistic distribution. The uniform

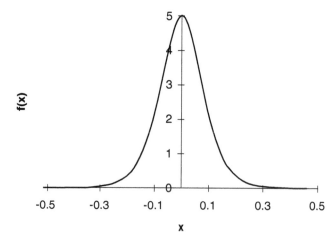

Figure 1.10. The logistic distribution

is totally flat, so consider modelling the distribution by adding a multiple, K, of the uniform to the logistic:

$$S(p) = \ln[p/(1 - p)] + Kp.$$

Figure 1.11 shows the shape of the PDF for this model and it is seen that it suitably combines the properties of the uniform and the logistic distributions, havin g a flattened logistic shape. The median is $K/2$ (= 0.1) indicating the shift in position of the distribution due to the added uniform distribution.

Example 1.18: In Example 1.15 the exponential and the reflected exponential distributions were combined. This was done with equal weightings for the two. Suppose now more weight is put on the right-hand tail and less on the left, as has already been done in Example 1.9, used for the previous illustration. Figure 1.5 illustrated the PDF for the model defined by the quantile function

$$S(p) = 0.8\ [-\ln(1 - p)] + 0.2\ [\ln(p)].$$

It will be seen that the distribution is now skewed to the right, since more weight is given to the exponential term than to the reflected exponential. We will call it the **skew logistic distribution**. The general basic form for this distribution can be written as

$$S(p) = \omega\ [-\ln(1 - p)] + (1 - \omega)[\ln(p)],\ 0 \leq \omega \leq 1.$$

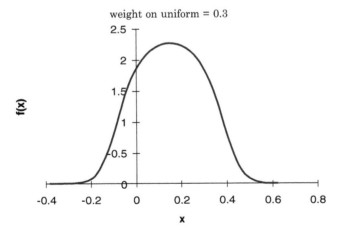

Figure 1.11. The uniform and logistic distribution

The parameter ω thus controls the relative weight given to the two tails of the distribution. A more useful alternative slightly reparameterizes this to give:

$$S(p) = \{(1 + \delta)/2\}[-\ln(1 - p)] + \{(1 - \delta)/2\}[\ln(p)], \; -1 \le \delta \le 1.$$

In this form, positive values of the parameter δ, the **skewness parameter**, correspond to distributions skewed to the right, as in the example above, where δ = 0.6. A zero value gives the symmetric distribution and negative values give a distribution skewed to the left (negative skewness). The change from the basic form to the general form is achieved by multiplying by a scale parameter and then adding a position parameter, so in general this is a three-parameter model. Each parameter corresponds to a visible feature of the distributional shape. Unlike the exponential and the symmetric logistic, this quantile function cannot be inverted to give x in terms of p. Thus there is no explicit CDF or PDF for this distribution.

Example 1.19: The previous examples have made use of the ability to add quantile distributions to generate new ones. If a quantile function is positive, it can be multiplied by another positive quantile function to obtain a new distribution. Consider, for example, the two positive distributions: the uniform distribution and the Pareto distribution. The Pareto is a distribution with a long tail to the right. It has a quantile function in basic form:

$$S(p) = 1/(1 - p)^\beta, \; \beta > 0.$$

The parameter β is a shape parameter and we denote the distribution by Pa or $Pa(\beta)$. The distributional range is, from $S(0)$ and $S(1)$, $(1,\infty)$. If the uniform quantile function, $S(p) = p$, is multiplied by that of the Pareto, the following quantile function is obtained:

$$S(p) = p/(1 - p)^\beta.$$

We will denote this by $U \times Pa$. The uniform distribution is a special case of the power distribution, defined by $S(p) = p^\alpha$, $\alpha > 0$. Thus $U \times Pa$ is a special case of the Power × Pareto distribution $Po \times Pa$:

$$S(p) = p^\alpha/(1 - p)^\beta.$$

Figure 1.12 shows its derivation and form. Notice that, like the skew logistic, this quantile function cannot be inverted to give an explicit CDF.

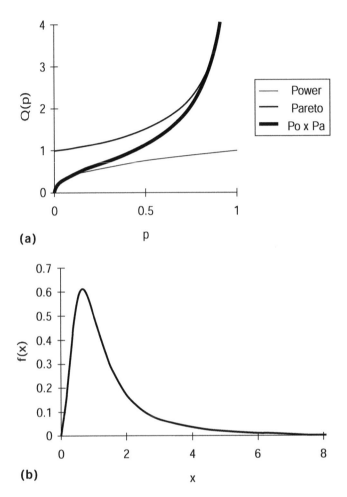

Figure 1.12. (a) The Power, Pareto and Power × Pareto distribution quantile functions. (b) The PDF for the Power–Pareto distribution

Example 1.20: As a final illustration of the quantile-based approach to modelling, let us return to the idea that models may contain both deterministic and random elements. In a report by Scarf (1991) the depths of corrosion pits in metal were modelled. The depth of the deepest pit on an experimental slab of metal was modelled by a distribution called the **generalized extreme value distribution**. This model will be discussed in Chapter 7. It has a quantile function

$$Q(p) = \lambda + \eta[1 - (-\ln(p))^\alpha]/\alpha.$$

The experimental data suggested that both the depth and its variability increased with time, t, in the same way and a power increase, t^γ, $\gamma > 0$, seemed appropriate. The model that combines both the required distributional and deterministic features is

$$Q(p) = \lambda\ t^\gamma + \eta\ t^\gamma[1 - (-\ln(p))^\alpha]/\alpha.$$

Here both position, $\lambda\ t^\gamma$, and scale, $\eta\ t^\gamma$, increase with the power of time from values of λ and η at $t = 1$. Models that express the distribution as a quantile function and involve the effects of other variables, in this case only one, time, are called **regression quantile models**. The value of the quantile form of the model becomes apparent if we wish to answer a question such as: How thick must the metal be to give a 99% chance of no corrosion holes going through the metal ($p = 0.99$) at time $t = 100$? Clearly, the metal must be thicker than the corresponding 99% quantile as given by the quantile regression model with $p = 0.99$ and $t = 100$.

We have shown through a set of examples that the quantile function can be used as the basis for a range of approaches to the construction of models for populations. The problem faced by the modeller is how to construct models that match the situation being studied. We have now seen some of the forms that the models might take and we have looked at ways of describing data. Before a match can be made, we need to look at summary ways of describing the features of the distributional models.

1.7   Simple properties of population quantile functions

We have seen so far pictures of several distributions defined as quantile functions. We have not as yet shown how to graph the probability density function of such distributions or how to describe their properties. So we need to look a little further at the quantile function and its behaviour.

The natural starting point is to substitute some values for p in $Q(p)$. This leads to population measures corresponding to those already used to describe samples.

$p = 0.5$	$Q(0.5)$	$= M,$	the Median.
$p = 0.25$	$Q(0.25)$	$= LQ,$	the Lower Quartile.
$p = 0.75$	$Q(0.75)$	$= UQ,$	the Upper Quartile.

These, together with the **distributional limits**, $Q(0)$ and $Q(1)$, give, in five numbers, a basic feel for the spread of the distribution over the axis.

Example 1.21: For the uniform distribution with $Q(p) = p$, the five values defined are $(0, 0.25, 0.5, 0.75, 1)$.

Example 1.22: For the unit exponential distribution, $S(p) = -\ln(1 - p)$:

$$S(0) = -\ln(1) = 0,\ LQ = -\ln(3/4),\quad M = -\ln(1/2) = \ln(2),$$

$$UQ = -\ln(1/4) = -\ln(1/2^2) = 2\ln(2) = 2M,\quad S(1) = -\ln(0) = \infty.$$

If a model has position and scale parameters, λ and η, then $Q(p) = \lambda + \eta S(p)$. The above measures of the shape of $S(p)$ are transformed in the same way to give the values for $Q(p)$. Thus, for example, for the median, using capital subscripts to refer to distributions (a natural notation that we will use in future),

$$M_Q = \lambda + \eta M_S.$$

It is often helpful to use measures of shape derived from LQ, M, and UQ. Thus as a measure of the spread of a set of data, we use the **interquartile range**,

$$IQR = UQ - LQ.$$

This range includes the central half of the probability.

The quantities $LQD = M - LQ$ and $UQD = UQ - M$ are the **lower** and **upper quartile deviations**, respectively. The difference between them gives a measure of skewness called the population **quartile difference**:

$$QD = UQD - LQD = LQ + UQ - 2M.$$

If we divide QD by the IQR, we obtain the **population Galton skewness coefficient**:

$$G = QD/IQR.$$

Notice that, as before relating $Q(p)$ to $S(p)$, by substitution

$$IQR_Q = \eta IQR_S \text{ and } G_Q = G_S.$$

Thus the IQR is independent of the position parameter and G is independent of both position and scale.

Example 1.23: As a further example, consider again the basic form of the skew logistic distribution:

$$Q(p) = \{(1 + \delta)/2\}[-\ln(1 - p)] + \{(1 - \delta)/2\}[\ln(p)], \quad -1 \le \delta \le 1.$$

The main shape measures are, on substitution,

$$p = 1/4. \ LQ = \{(1 + \delta)/2\}[-\ln(3/4)] + \{(1 - \delta)/2\}[\ln(1/4)]$$

$$= 0.0837\delta - 0.5493.$$

$$p = 1/2. \ M = \{(1 + \delta)/2\}[-\ln(1/2)] + \{(1 - \delta)/2\}[\ln(1/2)]$$

$$= \delta \ln(2).$$

$$p = 3/4. \ UQ = \{(1 + \delta)/2\}[-\ln(1/4)] + \{(1 - \delta)/2\}[\ln(3/4)]$$

$$= 0.0837\delta + 0.5493.$$

Using these in the derived statistics gives, on simplifying,

$IQR = \ln(3)$, $QD = -\delta\ln(3/4)$, and hence, evaluating the $\ln(.)$ numerically, $G = 0.26186 \ \delta$.

It is now more evident that δ is a skewness parameter.

The quantile properties of models derived by multiplying quantile functions together also have simple interrelations. Suppose we obtain $Q(p)$ by multiplication of $A(p)$ and $B(p)$, where $A(p)$ and $B(p)$ are non-negative quantile functions. Any quantile is obtained by multiplication and hence the median, for example, is $M_Q = M_A \times M_B$.

The quantile density function was introduced in Section 1.4. Another related quantity is obtained from the PDF, $f(x)$, by substituting for x with the quantile function, thus

$$f_p(p) = f(Q(p)).$$

This was called the **density quantile function** by Parzen (1979). To emphasize that it is basically the PDF expressed in terms of p and to avoid mixing up the quantile density and the density quantile,

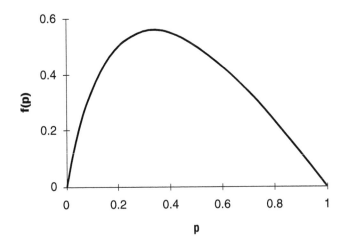

Figure 1.13. The p-PDF for the skew logistic distribution

the author will term this as the **p-PDF**. The shape of the p-PDF for the skew logistic, discussed previously, is illustrated in Figure 1.13.

The quantile density function and the p-PDF are closely related. As $x = Q(p)$ and $p = F(x)$ for any pair of values (x, p), it follows from the definition of differentiation that

$$(dx/dp)(dp/dx) = 1 \text{ so } (dQ(p)/dp)(dF(x)/dx) = 1,$$

hence $q(p)f(x) = 1$ and, therefore, expressing all in terms of p, $q(p)f_p(p) = 1$. The two functions $q(p)$ and $f_p(p)$ are thus reciprocals of each other. For example,

$$f_p(p) = 1 - p \text{ for the exponential distribution;}$$

$$f_p(p) = p(1 - p) \text{ for the logistic distribution;}$$

and

$$f_p(p) = (1 - p)^{1 + \beta/\beta} \text{ for the Pareto distribution.}$$

The previous result provides a way of plotting the PDF corresponding to quantile functions. If we let p take the values, say $p_i = 0.01$, $0.02, \ldots, 0.99$ and plot the points $(Q(p), 1/q(p))$, then we get the plot of points $(x, f(x))$, i.e., the plot of the PDF of x. Thus we can obtain plots of the PDF of a distribution entirely from the quantile function and its derivative, without having to be able to invert $Q(p)$ to get $F(x)$. The plots of PDF in this book are normally obtained in this fashion. The 99 points are best supplemented by additional tail probabilities. For example, p = 0.0005, 0.001, 0.002, etc., are usually adequate to get a good plot of the PDF into the tails. The p = 0 and p = 1 points may only be used if the distribution has finite limits for both $Q(p)$ and $f_p(p)$.

1.8   Elementary model components

Children's construction kits such as Lego and the classic Meccano involve two elements: a set of basic components and a means of joining them together. In Lego, these elements are both associated with the basic bricks. In Meccano, these are the flat metal shapes and the nuts and bolts for joining them. Section 1.6 considered the means of combining elements and illustrated them using a number of components. Section 1.7 defined some of the properties of importance in describing distributions and illustrated how the properties of constructed models can relate to those of their components. It is worthwhile now to list some of the components that comprise the modelling kit and look at their simplest properties. In doing this only the basic form will be considered, ignoring the position and scale parameters. There are no doubt many distributions that might form the kit for particular specialist areas. Here we just concentrate on a selection of the simplest ones that match familiar shapes across a breadth of statistical applications. Some of the component distributions have no parameters, others have a shape parameter. Table 1.2 gives the distributions and Table 1.3 lists some of their quantile properties.

Figure 1.14 illustrates the shapes of the p-PDF of some of these models, with $\beta = 0.5$ for the one-parameter distributions. Some of these models have been referred to previously. It is worth noting a few points about the tabulated information:

a.   As previously noted, the inverse uniform distribution is a special case of the Pareto when $\beta = 1$.

b.   The power distribution has a decreasing PDF only if $\beta > 1$.

c.   The shape parameter β has a complex influence on the form of the one-parameter distributions. The median and IQR will be modified by position and scale parameters. There will be a relation between the Galton skewness and the β parameter alone, but the relation will not be simple.

	Basic QF, $S(p)$	Distributional range
Parameter-free basic distributions		
Uniform distribution, U	p	$(0, 1)$
Inverse uniform, $1/U$	$1/(1 - p)$	$(1, \infty)$
Unit exponential, Exp	$-\ln(1 - p)$	$(0, \infty)$
Standard normal, N	$N(p)$	$(-\infty, \infty)$
One-parameter distributions $\beta > 0$		
Power distribution, Po	p^β	$(0, 1)$
Pareto distribution, Pa	$1/(1 - p)^\beta$	$(1, \infty)$
Weibull, W	$[-\ln(1 - p)]^\beta$	$(0, \infty)$

Table 1.2. Basic component distributions

	Median	Interquartile range	Quartile deviation
Parameter-free distributions			
Uniform distribution, U	0.5	0.5	0
Inverse uniform, $1/U$	2	2.66	4/3
Unit exponential, Exp	$\ln(2)$	$\ln(3)$	$\ln(4/3)$
Standard normal, N	0	1.35	0
One-parameter distributions			
Power distribution, Po	$1/2^\beta$	$(3^\beta - 1)/4^\beta$	$(1/4^\beta)$
			$(3^\beta + 1 - 2*2^\beta)$
Pareto distribution, Pa	2^β	$4^\beta(1 - 1/3^\beta)$	$4^\beta(1 + 1/3^\beta - 2/2^\beta)$
Weibull, W	$[\ln(2)]^\beta$	$[\ln(4)]^\beta - [\ln(4/3)]^\beta$	$[\ln(4/3)]^\beta + [\ln(4)]^\beta$ $- 2[\ln(2)]^\beta$

Table 1.3. Basic component distributions — properties

It will sometimes be helpful and insightful to have a notation to describe how the more common models are constructed from their components. The following form gives such a distribution built from two components: A and the reflected version of B, RB, using the additive skewed construction we have discussed previously in Example 1.18.

$$\text{Distribution} = (\lambda, \eta, \delta, A, +, RB).$$

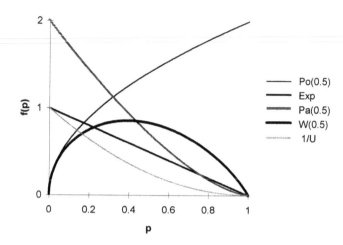

Figure 1.14. *p*-PDFs for some basic models

For example, the skew logistic is (λ, η, δ, Exp, +, RExp). Another example is the distribution (λ, η, δ, $Pa(\alpha)$, +, $RPa(\beta)$), which has five meaningful parameters. Classical distributional models have tended to keep to two or three parameters. It is, however, now clear that there are at least five features of a distribution that are expressed in models such as this, namely position, spread, skewness, right-tail shape and left-tail shape. Thus the use of multi-parameter models should not be regarded as ill advised. The issues are

    a.   How many distinct features are evident in the data and need to be modelled? Five distinct aspects need a model with five parameters.

    b.   Can we in practice match the different parameters in potential models to these features?

    c.   Have we sufficient data to carry out the detailed matching? If we have only small data sets we may be forced to use simpler models.

    d.   One general principle is always to use the model with the minimum number of parameters needed to adequately model the data. Putting in too many parameters will lead to a better fit to the actual data obtained but probably a worse fit to future data, since the model is picking up essentially random features of the current data.

Distributions based on addition but without a skewness parameter can be described in basic form as, for example, $S(p) = A + RB$. For multiplicative models the distributions can be multiplied, providing the variables are positive, which is true for nearly the entire previous list of components. Thus for the basic form, $S(p)=A \times B$. For example, the Uniform $\times$ Pareto, previously introduced, is $U \times Pa(\beta)$.

It is evident that with the few simple components now introduced there are a large number of possibilities for different shapes of distributions. One can now begin to see the development of distributional models as a process of model construction, using a kit consisting of a variety of components and a variety of ways of joining them together.

1.9   Choosing a model

The technical term for choosing a suitable model is **identification**. This topic is discussed later in the book, in Chapter 8, where the aim is to build models to get the most appropriate shapes for the data. For the purpose of introduction, this process is reduced to trying out several potential models in order to choose the best. In practical situations the modeller rarely starts with total ignorance. Even a little experience in an application area will suggest that there are certain models that commonly occur. The graphical studies of the sample data will give a feel for the basic shape of the distribution. These can be matched with the forms that are already familiar or may guide in the construction of new models using some of the ideas introduced in Section 1.6. Thus it is assumed, for now, that several candidate models can be compared with the data in a visual fashion.

The easiest way to do this comparison is via what is commonly called a **Q-Q plot**. Suppose $Q(p)$ is a suitable model, $x_{(1)}, x_{(2)}, x_{(3)}, \ldots, x_{(n)}$ are the n-ordered observations and p_r, $r = 1, 2, 3, \ldots, n$, are the corresponding probabilities; $p_r = (r - 0.5)/n$. The Q-Q plot is a plot of the points $(Q(p_r), x_{(r)})$, i.e., the n data quantiles, $x_{(r)}$, against the corresponding model quantiles, $Q(p_r)$. This plot will give, for a "good" model, an approximately straight line since the model gives the p-quantiles, x_p, as a function of p; $x_p = Q(p)$. The line should be at 45°. In practice, $Q(p)$ is often a model "fitted" to the set of data, denoted by $\hat{Q}(p)$. Hence the plot of $(\hat{Q}(p_r), x_{(r)})$ is called the **fit-observation diagram** and the 45° line the **line of perfect fit**. If we use just a basic model the plot will still be approximately linear but the intercept and slope of the line will depend on the unknown position and scale parameters of the distribution. From the point of view of choosing the basic model these parameters do not matter, so interest simply focuses on the occurrence of an acceptably straight line. An inappropriate model will show some systematic curvature. A

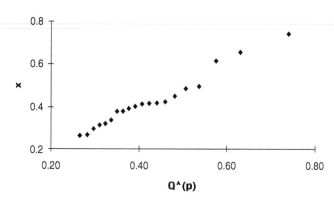

Figure 1.15. Flood data — Fit-observation plot for a Weibull distribution

feature of such plots is that even the best always show some "snakelike" wandering, but this can usually be distinguished from the systematic, model-generated features of an inappropriate model.

Figure 1.15 shows the fit-observation diagram for a model used with the flood data of Example 1.1. The plot indicates a good choice of model. It is also possible to plot together the sample values of Dx/Dp and the corresponding values of the model quantile density function, $q(p)$, which is the population dx/dp. To illustrate, consider two potential models for the flood data. For the moment it does not matter particularly what these models are. In Figure 1.16 they are called EV and Weib. The plot gives the shapes of $q(p)$ for these two models; also shown is the sample $q(p)$, which is the data plot of Dx/Dp. It is evident that both models show the same basic shape as the data. It is also evident that the right-hand tail of the EV model shows a steeper slope than the Weib model.

Obviously, ways of choosing the best model need to be studied further. However, it is always important to refer back to the data. It is clear here that with such a small and variable set of data, both models will provide a passable description. It may be that the practical application will influence the choice of model. For example, if our objective is to use the right tail of the model in the design of flood defences, then the EV model will evidently give more conservative, cautious values than the Weib model. The plot also raises issues about whether or not we could get better data, e.g., say annual data, to help discriminate more clearly between the two models.

For models with unknown position and scale parameters, the straightness of the line of the Q-Q plots can still be examined. For

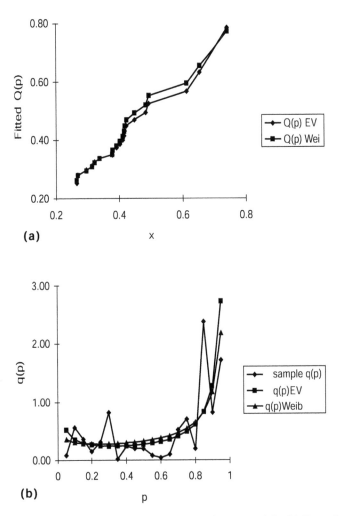

Figure 1.16. Flood data — (a) Observation-fit plots for two models; (b) Quantile density plots for two models and the data

unknown shape parameters, there will be a need to give numerical values to the parameters to enable the Q-Q plot to be drawn. For the models used in the example it is necessary to settle tentatively on a model and use the methods in the next section to allocate numerical values. Thus the modelling process is seen not to be a linear one, but often an iterative process of revising first thoughts in the light of later information.

1.10   Fitting a model

It is now assumed that there are a set of data and a suitable form of model. The model will probably involve position and scale parameters and may even have several more. In much of statistics there has been a tendency to keep to models with two or three parameters. However, when large data sets are considered, their properties can at least be listed as position, scale, skewness, right-tail shape and left-tail shape, i.e., five distinct features that may need five distinct parameters. Numerical values for these parameters are required so that the Q-Q plot is a good line at 45°. Thus it is required to **fit** the model to the data. The process of giving numerical values to parameters is called **estimation**. Almost all the literature on estimation is based on defining distributions by their probability density functions or, very occasionally, by their cumulative distribution functions. If these do not have explicit form but do have explicit quantile functions, we are forced to consider how to estimate the parameters in the quantile function. It is often the case that having fitted a distribution we then use the quantile form of the model in the application, for example, to assess some high quantile value like $Q(0.95)$ or $Q(0.999)$. In these situations it is natural to ask whether or not the quantile function might here be used as the form to be fitted.

There are a vast number of methods of estimation. In Chapter 9 we will look in detail at those most suitable for fitting quantile models. For now we look at two methods that are both simple and appropriate. The simplest is commonly called the **method of percentiles**, although it has sometimes been called the method of quantiles (e.g., see Hald (1998) and Bury (1975)). The p_i-th percentile of the population described by the distribution $Q(p)$ is simply $Q(p_i)$, where $100p_i$-th is a suitable percentage. The corresponding sample percentile, quantile, can be obtained from the data we saw in Section 1.3. The method of percentiles chooses a set of k-suitable percentile points, $p_1, ...,p_k$, where k is the number of model parameters. The parameters are then given the numerical values that ensure that at these k specific points the percentiles of the sample and fitted population are identical. Thus the model fits perfectly at these points.

> **Example 1.24:** For the exponential distribution, the position parameter is zero, the left-hand end of the distributional range, and the scale parameter is the unknown η. The population median, the 50% percentile, is given by $Q(0.5) = \eta \ln 2$, the sample median is given by m. The method of percentiles chooses η to equalize these two, so the estimated η, denoted

by $\hat{\eta}$, is given by $\hat{\eta}$ = m/ln2. Thus a sample median of 4.3 leads to an estimated η of 6.2. The **fitted model**, denoted by $\hat{Q}(p)$, is thus $\hat{Q}(p)$ = -6.2 ln(1 − p). The method describes what to do but it does not specify what percentiles to use for the equalisation. The median was used here as a reasonable central value. If, however, the fitted model is to be used to make predictions about relatively rare events, with high values of p, it would make more sense to match population and sample for p values out towards the right tail. However, there is a need for caution since the sample percentile will usually have a higher natural variability out in the tail, especially if the sample size is small. Only if the sample size was large would one use, say, the 90th percentile. This illustrates the points that at least some methods of estimation involve matters of judgement and that in all fitting one needs to take into account the purpose for which the fitted model is required.

Example 1.25: Consider the symmetric logistic distribution:

$$Q(p) = \lambda + \eta \ln[p/(1 - p)].$$

Three symmetrically placed percentiles are LQ, M 46$q$and UQ. As we only have two parameters, we only need two values. However, consider the nature of the parameters. The parameter λ is a position parameter which suggests the use of the median. In fact, $M = Q(0.5) = \lambda$. Thus λ is estimated by the sample median, so $\hat{\lambda} = m$. The parameter η is a scale parameter and a measure of scale is provided by the interquartile range. The population value of the IQR is

$$IQR = UQ - LQ = \eta[\ln(3) - \ln(1/3)]$$

$$= 2\,\eta\ln(3).$$

If the sample interquartile range is iqr, then an estimate of η, given by matching population and sample values, is $\hat{\eta} = iqr/2\ln(3)$. Thus rather than matching at points we are matching quantile properties.

   In the discussion of identification, the straightness of the fit-observation plot was used to choose a good model. This same criterion can be used as a basis for choosing the parameter values. The parameters can be chosen to give a straight line on the fit-observation plot that closely fits the data points. Thus let θ represent possible values of the parameters of the quantile function and $Q(p;\theta)$ the chosen form. The deviations of the observations from a given quantile function, called here the **distributional residuals**, are $e_r = x_{(r)} - Q(p_r;\theta)$, for r = 1, 2, ...,n. These should be small for a good fit, i.e., a good choice of θ.

The classic approach to fitting straight lines is the **method of least squares**. This method is based on choosing the parameter estimates to minimize the sum of the squares of the residuals, Σe_r^2. The sum of squares (SS) gives one measure of the **discrepancy** between the ordered observations and their fitted values. An alternative measure, the first historically, is based on $\Sigma |e_r|$, where the $|\cdot|$ denotes the absolute value, which in many computer languages is ABS(.). The sum of absolute distributional residuals is one that is natural to use for quantile distributions. The discrepancy is just based on the numerical distances of the ordered observations from some measure of their position as given by the fitted distribution. There is thus a **method of least absolutes** similar to the method of least squares. For least squares, the appropriate line of fit is based on the means of the distributions of the ordered variables $X_{(r)}$. For the least absolutes it should be based on the population medians of these distributions. For the value $p_r = (r - 0.5)/n$, the value of $Q(p_r)$ is an approximation to the mean of the distribution of $X_{(r)}$. Thus the least squares criterion for fitting distributions is (approximately):

$$\text{Choose } \underline{\theta} \text{ to minimise } \Sigma(x_{(r)} - Q(p_r; \underline{\theta}))^2$$

We will show later that we can calculate a probability p_r^* such that $Q(p_r^*, \underline{\theta})$ is exactly the population median of $X_{(r)}$ for all r and all reasonably behaved continuous distributions. We will show later that for minimising absolute values the natural basis for a discrepancy measure is the median rather than the mean. Thus the criterion and method in this case are exactly

$$\text{Choose } \underline{\theta} \text{ to minimize } \Sigma | x_{(r)} - Q(p_r^*; \underline{\theta})| .$$

This approach is not quite the traditional least absolutes (or least squares) since $x_{(r)}$ are ordered data and the model is the distributional model defined by its quantile function. To emphasize this difference we call these approaches the **methods of distributional least squares/absolutes**, DLS and DLA. We use e_r the distributional residual appropriate to whichever method we are using. Table 1.4 shows the steps in carrying out these methods of estimation. Notice that the steps are quite general and ultimately depend on choosing the parameters to minimise a criterion calculated from the model, the initial parameter values and the data. Many programmes, including standard spreadsheets like Excel®, provide this general numerical

Step	Objective	Calculation
1	Define distribution	$Q(p;\underline{\theta})$
2	Set initial value(s) of $\underline{\theta}$	$\underline{\theta}_0$
3	Calculate p	$p_r = (r - 0.5)/n$ for DLS,
		p_r^* (Section 4.2) for DLA
4	Calculate fitted QF	$Q(p_r; \theta_0)$ DLS or $Q(p_r^*;\theta_0)$ for DLA
5	Calculate criteria C	$\Sigma \mid (x_{(r)} - Q(p_r ;\theta_0))^2$ for DLS
		$\Sigma \mid x_{(r)} - Q(p_r^* ;\theta_0) \mid$ for DLA
6	Choose $\theta_0 = \hat{\theta}$	using standard software
	to minimize C	
7	Plot fit-observation	$(Q(p_r$ or $p_r^*;\hat{\theta}), x_{(r)})$
	diagram	

Table 1.4. Steps for fitting with distributional least squares/absolutes

minimization facility (called Solver® in Excel®, which is a registered trademark of the Microsoft Corporation). Although there are many powerful statistical packages now available, it should be noted that there are no significant calculations in this book that cannot be done on a spreadsheet.

Table 1.4 is the first of a number of tables in this book that give the steps, the algorithm, for a calculation. It is hoped that these, together with "layout" tables, such as Table 1.1, will give the readers sufficient guidance that they can readily develop examples using their own data, which is ultimately the only way to learn.

Least squares has been almost universally used for fitting lines to data, as it is computationally simple and has good statistical properties. There is a vast literature stretching back over 200 years dealing with the fitting of deterministic models. The use of least squares as distributional least squares dates back only a relatively short time. We will see that the method of least absolutes provides an exact, robust and universal approach to estimating distributional parameters for models based on quantile functions. It will be used and explored in future chapters. The literature on these topics is very concerned with examining special cases, obtaining formulae appropriate for each distribution. In an age of readily available computing power, the criteria one uses should be based on the relevance to the application and appropriateness to the statistical objectives and not just traditional computational ease. Having determined the criterion and the distribution, the model can be fitted by the steps of Table 1.4.

Weibull $Q(p) = \lambda - \eta[\ln(1-p)]^\beta$ estimates $\hat{\lambda}$ = 0.251

$n = 20$ $\hat{\eta}$ = 0.193

distributional residual = e $\hat{\beta}$ = 0.711

For future plots   Sum $ABS(e)$ = 0.323

r	x	p	$\hat{Q}(p)$	$ABS(e)$	
1	0.265	0.025	0.265	0.000	$\hat{q}(p)$
2	0.269	0.075	0.282	0.013	0.341
3	0.297	0.125	0.297	0.000	0.292
4	0.315	0.175	0.311	0.004	0.273
5	0.323	0.225	0.324	0.001	0.265
6	0.338	0.275	0.337	0.001	0.263
7	0.379	0.325	0.350	0.029	0.264
			etc.	38	38

Table 1.5. Layout for fitting by distributional least absolutes

Example 1.26: To illustrate the method of distributional least absolutes
Table 1.5 shows the layout of the calculation for fitting the flood data of
Table 1.1 using the method of distributional least absolutes. The Weibull
distribution is used as the distribution to illustrate the fitting, this being
one of our set of simple component models. For the flood data it is evident
that we need a position parameter, which is often set to zero, but the flood
data will be well above zero, and also a scale parameter. The model is thus

$$Q(p) = \lambda + \eta[-\ln(1-p)]^\beta.$$

Following the steps of Table 1.4, the Weibull model was defined initially
with guestimated parameters, $\hat{\theta}$, as $\hat{\lambda} = 0.2$, $\hat{\eta} = 0.2$ and $\hat{\beta} = 0.5$, based
on a little exploration of values. The values of p_r are, for simplicity of
illustration, those used in the previous study of this data. Hence $Q(p_r;
\hat{\theta})$ can be calculated for each observation. Thus it is now possible to
calculate the distributional residuals and hence their sum of absolute
values, shown at the top of the table. The three parameters are then
adjusted to obtain the minimum value of the criterion. The table and
Figure 1.13 show the resulting values and fit-observation plot for the
fitted Weibull distribution. The fitted Weibull is thus

$$\hat{Q}(p) = 0.251 + 0.193[-\ln(1-p)]^{0.711}$$

The fit-observation plot of the data for this model shows a reasonable linearity. One beauty of the method is its simplicity in choosing the parameters to provide the best plot for the visual assessment of the fitted model. The method is not without its weaknesses, as we shall see. However, it is straightforward and effective and we will illustrate its wide applicability. Table 1.4 shows the basic form of the calculations behind Figure 1.14.

1.11   Validating a model

The daily practice of statistics is not a series of unique problems. In many areas, such as quality control, the same form of data arises repeatedly. We are faced with data for which a model is already proposed. In such situations we will want to be assured that the model is still appropriate for the new data. If it is not, we will need to know whether the model is now totally inappropriate, or, for example, it is just the position parameter that has shifted a little. The process of model checking is called **validation**. We will systematically cover this topic in a later chapter. For now it is sufficient to note that there are two naturally linked approaches: one graphical, one mathematical. It is evident, for example, that if further maximum flow data was available to add to that of Table 1.1, then the previous fitted model of Example 1.26 could be used in a fit-observation diagram constructed with the new data. If the situation has changed, the new data will not show a good fit. The distributional residuals for the new data, using the old $\hat{Q}(p)$, could be plotted with those of the old data. If they merged, the validation will have given confidence in the continued use of the fitted model. In more formal methods the objective might be to test whether or not the position parameter changed for the new data. A possible change in position could be studied at a very simple level by comparing the median of the new data with the median of the fitted model. This would require a study of the behaviour of the sample median; however, the logic of the validation is clear.

1.12   Applications

Chapter 14 will discuss a number of areas of the application of statistics where quantile-based distributions are of particular use. These applications, on occasion, lead to specific developments of the quantile modelling approach. For the purpose of this overview we consider just one application area and that only in very general terms. Statistical quality control is concerned with the study of the variability that occurs in manufacturing and service processes. Specific concerns are

1. The detection of possible changes in the process from its specified behaviour. The classic example of this is where a quantity, v, calculated from a process sample, has a known normal distribution, $Q(p) = \lambda + \sigma N(p)$, where $N(p)$ is the quantile function of the standard (basic) normal distribution when the process is under proper control. The most common loss of control in a process is due to a change in λ away from its target value of μ. A simple device for monitoring for such a change is a graph, called a **control chart** on which sample values of v are plotted in sequence. On the same graph are plotted lines at $L = \mu + \sigma N(0.00135)$ and $U = \mu + \sigma N(0.99865)$, called the **action limits**. These correspond to $Q(p) = \mu \pm 3\sigma$. Clearly the chance of v lying outside these lines for any given sample is 0.0027. This probability is sufficiently small that the occurrence of points outside the action limits casts doubt on the appropriateness of the model and, specifically, on the value of λ. Thus this control chart provides one way of continuously monitoring the process. The use of the quantile function in the formulation immediately shows how to construct action limits for any distributional model, simply replacing $N(p)$ by the appropriate $S(p)$.

2. The procedure of (1) seeks to monitor short-term problems that may occur in a process. There is also concern for improving the inherent quality of a process. If the process has a target value for an output, the quality may be measured by the variability, as indicated by η. For some other process, the quality may be indicated by the scale or shape parameter of the distribution of, say, the % of an impurity or the concentration of the desired chemical output. There are also situations where weakness in the process operation shows itself by skewness in the quality measure(s). The focus in all these situations is on investigating how to modify the process to produce 'quality improvements', i.e., decreases or increases in the appropriate parameters.

3. As in all areas of application, the statistical methods used have to recognize practical considerations as well as theoretical ones. Historically, in process control this has meant the use of simple measures rather than complex ones and the wide use of graphical techniques. Both of these benefit from the use of quantiles and their properties.

1.13 Conclusions

The purpose of this chapter has been to give the reader an overall view of the 'wood' before embarking on a more detailed look at its sections and some of the 'trees'. We have of necessity had to select illustrative 'trees', which later will have to be looked at again, as part of the more detailed look at their particular section of the woods. The prime consideration has been to show that the use of the quantile functions provides the modeller with both a practical, useful way of looking at distributional data and a kit of simple components that can be used to construct models of the complexities of real data. We have in the process of our discussion taken an overall look at the modelling process. The process has five main stages, although, as we have seen, we do not necessarily go through these in sequence but often repeat steps after revising thoughts in the light of our modelling experience. Our models are forever tentative and we will always be seeking to improve them. The five stages are

The development of a construction kit — This involves the modeller in:

Getting to know the most useful components and their properties.

Knowing different ways of joining the components together to achieve desired ends.

Developing model components relevant to the particular application area.

Playing with the kit, which is a good way of starting. The problems at the ends of later chapters are there to provide this opportunity.

The identification of suitable models for the data — In the past, identification has revolved around finding suitable models within the library of standard models. With the construction kit approach there is the additional possibility of identifying the properties of different elements of the data, e.g., tail shapes and skewness, and constructing new models from components to reflect these properties.

The fitting of models — All the common methods of fitting, estimation, can be used for models expressed in quantile form, but as we have seen, some methods have particular suitability.

The validation of fitted models — Here there are a variety of graphical methods, such as the fit-observation plot, and later we will consider more analytical approaches.

Application — It is sufficient here to note that the quantile view of models is often suitable to particular applications. Further, the modelling of the deterministic part of models in many applications is based on the use of appropriate construction kits. Thus the total modelling exercise takes on a coherency when both deterministic and random components can be used to build the model.

We now have an overall process of background construction, identification, estimation, validation, application (and iteration). The rest of this introductory text now looks in more detail at each of these elements and at the statistical methods required for them.

CHAPTER 2

Describing a Sample

2.1 Introduction

Chapter 1 gave an overview of the subject of this book. We now begin to look at the subject in a more comprehensive fashion. We will of necessity look again at some of the material of that chapter and we start with the same central topic of statistics, which is the study of data.

We are surrounded by data that shows an inherent variability. The issue addressed in this chapter is how we may describe this variability. Traditional statistics focuses on the proportions of observations lying in particular sections of the range of possible values. Corresponding plots and measurements focus on this theme, for example, the classical histogram. When we look at a single observed value, 235.1 say, it tells us very little on its own. For it to be meaningful, we need to know how it relates to the rest of the data. If we are told that all the data lie between 230 and 240 and that the observation is 12th out of 40 in increasing order, then we begin to get more of a picture of the situation. In the approach of this book we will be emphasising the ordered positions of data and the proportions of the data lying on either side of individual observations. This view draws to the fore a further set of plots and measures that add to the understanding of the data. They do not replace the others; they supplement them, giving a broader view of the data. Experience has always taught that the more ways one looks at a set of data the greater the understanding provided, and conversely the less likely the erroneous deduction.

Data in the form of counts, referred to as observations on a **discrete variable**, play only a minor role in our discussions and are briefly dealt with in their own section. For the rest of the book we focus on data with a continuous measurement scale, such as weights and lengths, which are observations on a **continuous variable**.

2.2   Quantiles and moments

In Chapter 1 we looked briefly at measures of the position of observations within the whole data set and introduced quartiles, quantiles,
and functions based on them. We now need to define these quantities
carefully. We start by looking at the **sample** or **empirical cumulative distribution function**, $\tilde{F}(x)$. We define this by

$$\tilde{F}(x) = \text{Proportion of sample} \leq x.$$

For a sample of n observations this can only take the values 0, $1/n$,
$2/n$, ..., 1. It thus forms a step function as shown in Figure 2.1(a). The
dots on the lines show that, for example, $x_{(2)}$ gives the value $2/n$. The
sample or **empirical quantile function**, denoted by $\tilde{Q}(p)$, is the
inverse of this, i.e., it is the plot with the axes interchanged as in
Figure 2.1(b). Formally we have

$$\tilde{Q}(p) = x_{(r)}, \ (r-1)/n < p \leq r/n$$

Again we have a step function, which is correct for a finite set of
data and appropriate for a discrete variable that can only take on
specific values of x. Our interest, however, is in observations on a
continuous variable. Thus if we only had sufficient data the plot
would begin to look like a continuous curve. To model this more
appropriately with our finite sample, we join together the mid-points
of the steps, as in Figure 2.1(c), to create a continuous curve made
of line segments. Notice that we cannot draw the curve outside $(1/n,
(n-1)/n)$ and that the p value corresponding to $x_{(r)}$ is $p_r = (r-0.5)/n$,
$r = 1, 2, ..., n$, as was used in Chapter 1. Thus for the integer values
of r we have

$$\tilde{Q}(p_r) = x_{(r)}.$$

For p in general, we let $r = np + 0.5$, so r may be fractional, and g
$= r - [r]$, where $[r]$ denotes the integer part of r. It will be seen in
Figure 2.1(d), where for simplicity we put $[r] = s$, that $\tilde{Q}(p)$ lies between
$x_{(s)}$ and $x_{(s+1)}$ in the proportion g and $1 - g$. Thus we can write for most
p and corresponding r:

$$\tilde{Q}(p) = (1-g)x_{(s)} + gx_{(s+1)}, \text{ for } p \text{ in } (1/n, (n-1)/n).$$

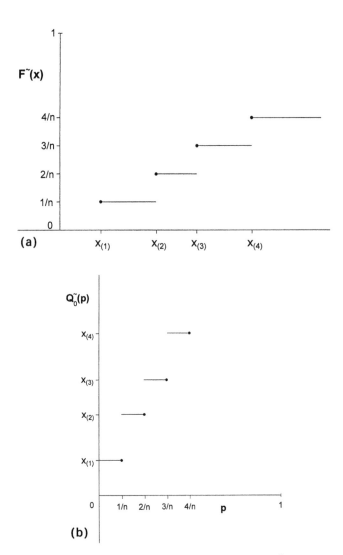

Figure 2.1. (a) The empirical cumulative distribution function, $\tilde{F}(x)$; (b) The empirical quantile function, discrete form, $\tilde{Q}_o(p)$

This gives the form we will use for the sample quantile function for continuous variables.

Example 2.1: Consider the seven observations 1,3,5,5,6,7,8.

For the median $p = 0.5$, so $r = 4$ and $g = 0$, hence $m = x_{(4)} = 5$.

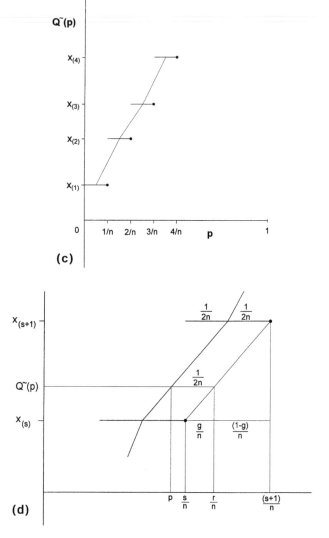

Figure 2.1. (continued)   (c) The empirical quantile function, a continuous form, $\tilde{Q}(p)$;
(d) Interpolation

For the lower quartile $p = 1/4$, so $r = 9/4$, $[r] = 2$, and $g = 1/4$, hence lq
$= 3 * 1/4 + 5 * 3/4 = 4.5$.

For the upper quartile $p = 3/4$, so $r = 23/4$, $[r] = 5$, and $g = 3/4$, hence
$uq = 6 * 3/4 + 7 * 1/4 = 6.25$.

Thus the interquartile range is 1.75.

It sometimes happens that the data are only available in summary form in a **frequency table**, as Table 2.1 illustrates. The original data are summarized in the table by giving the number, i.e., the frequency, of observations lying in specified ranges called **class intervals**. Notice that the interval given by, say, 11^+ to 13, includes the exact value 13 but starts just beyond the value of 11, which would be classified in the previous interval. The sum of the frequencies gives the total number of observations. To evaluate the medians and quartiles in these circumstances requires some manipulation of the table. As the use of frequency tables usually implies large data sets, we will not make the type of distinction we made between $\tilde{Q}_o(p)$ and $\tilde{Q}(p)$, which becomes negligible for large n.

Table 2.1 is created so that the cumulative proportion p represents the proportion of coils with weights less than or equal to the value of x. Figure 2.2(a) gives the quantile plot by plotting the interval end points against the proportions. We have joined the known values by line sections, a not unreasonable procedure (provided, that is, that our measurements are on a continuous variable). The median is the value with $p = 0.5$. This lies in the cell 19 to 21. If it is assumed that the observations lie uniformly in the cell, then the proportion of the frequencies needed to get to the median from $x = 19$ must correspond to the proportion of the additional p to get from 0.37 to 0.50. Thus

$$\frac{m - 19}{21 - 19} = \frac{0.50 - 0.37}{0.79 - 0.37}$$

Hence $m = 19.62$. A similar calculation for the lower quartile gives $lq = 17.59$. The proportion of observations less than or equal to 15 is 0.07 hence the 0.07 quantile is 15. Suppose, however, that we need the 0.9 quantile. By inspection of the table this will lie in the 21^+ to 23 class interval which has boundary proportions 0.79 and 0.99. Thus comparing corresponding proportions for $Q(0.9)$ and $p = 0.9$ gives

$$\frac{\tilde{Q}(0.9) - 21}{23 - 21} = \frac{0.90 - 0.79}{0.99 - 0.79}$$

Hence the 0.9 quantile, $\tilde{Q}(p)$, for the data is 22.1. This may also be referred to as the 90th percentile or the 9/10 fractile. Figure 2.2(b) also shows the sample p-PDF obtained as in Table 2.1 by plotting Dp/Dx (change in p across interval over class width) against the mid-value of p in the interval.

Weight interval	Frequency	$\tilde{Q}(p)$	Cumulative frequency	Cumulative proportion p	Dp	$Dx = 2$ Dp/Dx	Mid-p	Mid-x
−11	0	11	0	0.0				
11^+–13	3	13	3	0.03	0.03	0.015	0.015	12
13^+–15	4	15	7	0.07	0.04	0.02	0.05	14
15^+–17	13	17	20	0.20	0.13	0.065	0.135	16
17^+–19	17	19	37	0.37	0.17	0.085	0.285	18
19^+–21	42	21	79	0.79	0.42	0.21	0.58	20
21^+–23	20	23	99	0.99	0.20	0.1	0.89	22
23^+–25	1	25	100	1.00	0.01	0.05	0.995	24

Table 2.1. The frequencies of 100 metal coils with given weights as ordered

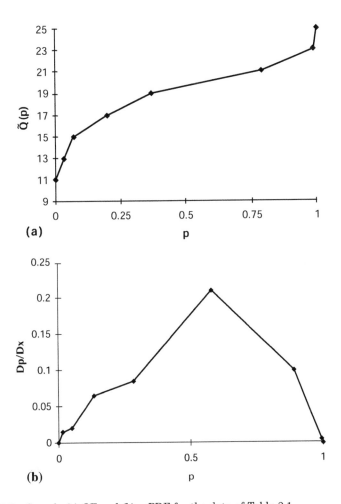

Figure 2.2. Sample (a) QF and (b) p-PDF for the data of Table 2.1

Notice the reversal that takes place in moving from raw data to a frequency table. For the raw data, the observations exhibit random behaviour and the proportions are set in the calculations. When a frequency table is used, the values of the variable are defined by the boundaries of the intervals and it is the proportions that exhibit the random behaviour. Notice also that we have had to make assumptions about how the data lie in the intervals. We do not know, so our calculations are approximations. Creating a frequency table loses information about the detail of the data. It may help in showing the big picture, but care needs to be taken with data in this form. In general, we will seek to emphasize methods that make use of the raw data rather than data in frequency tables. Historically, frequency tables were usually used because of the difficulty in doing calculations and handling graphics with large numbers of observations. In the days of the computer this ceases to be a major problem.

The median, as the middle value, gives a numerical indication of the position on the axis of a set of data. The other common measure of the position is the **sample mean** or **average**. This is the sum of the values divided by their number. For a set of data it is defined by

$$\bar{x} = \Sigma \, x_i/n,$$

or for frequency table data by

$$\bar{x} = \Sigma \, f_i \, x_i/n,$$

where x_i, in this case, is the mid-value of the interval with frequency f_i.

2.3   The five-number summary and measures of spread

The median and quartiles have been presented as means of indicating some aspects of the values taken by a set of data. Two other values that are helpful in summarizing a set of data are the smallest observation, s, and largest, l. We thus have a five-number summary of a set of data provided by s, lq, m, uq, l. For a very large set of data it is best to let a computer spreadsheet or a statistical programme find the values. For data sets up to, say, 100 observations a simple presentation called a stem-and-leaf plot is useful, as Table 2.2 illustrates. The data are simplified into "tens," which form the "stem," and units, which are ordered as the "leaves."

The five-number summary for the $n = 50$ observations can be read directly from the plot by counting. The median is the average of the middle two observations and the quartiles are the 13th and 38th observations. Hence

$$s = 270, \; lq = 550, \; m = 670, \; uq = 840, \; l = 1960.$$

To construct a stem-and-leaf plot, it is best to construct it with the data in its natural order in the leaves and as a second stage put

Stem hundreds	Leaf tens
2	7
3	4 6 7
4	4 5 5
5	0 0 2 3 4 5 7 8
6	0 0 2 2 2 3 3 5 6 7:7 9 9
7	0 0 2 4 5 8
8	1 2 4 4 6 6 7 7
9	5
10	0
11	0 1 6 8
12	
13	
14	5
15	
16	
17	
18	
19	6

The data are a set of the weights of 50 babies, rounded to the nearest ten, thus 677 is presented as a 6 in the stem and an 8 in the leaf.

Table 2.2. A stem-and-leaf plot

the leaves in order. Figure 2.3 shows a diagram called a box plot that gives a visual presentation of the five-number summary. The five values provide the five horizontal lines against the scale. The central box contains half the data values split at the median. The "whiskers" go out to the largest and smallest observations, or if there is a large data set, they may go to some points beyond which plotted observations are counted as outliers; $(lq - 3iqr/2, uq + 3iqr/2)$ provide such whiskers.

An important attribute of any set of data is its spread or variability. There are five useful measures of spread: the range, the interquartile range, the mean absolute deviation, the median absolute deviation, and the standard deviation. The first two come directly from the five-number summary, the **range**, w, being the difference between the largest and smallest values, and the **interquartile range**, iqr, being the difference between the quartiles. Thus

$$w = l - s, \; iqr = uq - lq,$$

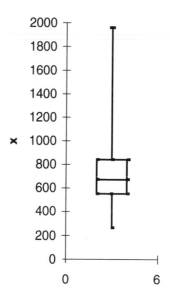

Figure 2.3. A box plot for the data of Table 2.2

The **mean absolute deviation** and **standard deviation** both make use of the deviations, d_i, between the observations and their average. Thus

$$d_i = x_i - \bar{x}.$$

The magnitudes of the deviations clearly indicate the variability of the data. We cannot use their ordinary average as a measure of spread as they sum to zero. We therefore either take their numerical, absolute values, denoted by $|d_i|$, or their squares to remove the cancelling negative signs. We thus have

the **mean absolute deviation**, MAD $= \Sigma \, |d_i|/n$

the **median absolute deviation**, MedAD $=$ sample median of $|d_i|$

and

the **sample variance**, $s^2 = \Sigma \, d_i^2/(n-1)$.

The use of the devisor $(n-1)$ in place of n is a technical modification. The measure of spread s is the **sample standard deviation**. Consider as a simple example the values 5, 5, 7, 8, 10, which have an average of 7. Thus the deviations are –2, –2, 0, 1, 3, which as we have noted sum to zero. The absolute deviations are 2, 2, 0, 1, 3 so MAD = 8/5 = 1.6. MedAD is 2. The squared deviations are 4, 4, 0, 1, 9. Hence $s^2 = 18/4 = 4.5$ and $s = 1.12$. Notice that if, for example, 10 was mistyped as 100, s^2 is most altered, MAD less so, and MedAD is unaltered. Thus s^2 is a sensitive measure, MAD is said to be more robust and MedAD is highly robust.

2.4   Measures of skewness

An important feature of any set of data is its symmetry or lack of it as indicated by its skewness. As for measures of position and spread we will look at measures of skewness based on both quantiles and moments.

As we have seen, a simple device for looking at skewness depends on using the deviations from the median. We define the **sample upper quartile deviation**, *uqd*, and the **sample lower quartile deviation**, *lqd*, by

$$uqd = uq - m, \quad lqd = m - lq.$$

The difference between these is the **sample quartile difference**, *qd*, where

$$qd = uqd - lqd.$$

For a perfectly symmetrical distribution of observations $qd = 0$, *qd* large and positive indicates a long tail to the right, a positive skewness, and vice versa. To make this measure independent of the scaling of the numbers involved we divide by the interquartile range, as a measure of spread, giving

$$g = qd/iqr = (uqd - lqd) \, / \, (uqd + lqd).$$

This is Galton's skewness index for the sample. Note that as all its elements are positive it lies in the range (–1, 1).

If we repeat the above logic replacing the quartile fraction 1/4 by some general proportion $p(0 \le p \le 0.5)$, we can define a **lower p-deviation** by

$$ld(p) = m - \tilde{Q}(p)$$

If we denote $1 - p = q$, then we also have an **upper p-deviation**

$$ud(p) = \tilde{Q}(q) - m,$$

For practical purposes we use $\tilde{Q}(p_r) = x_{(r)}$ and $\tilde{Q}(q_r) = x_{(n+1-r)}$ with $p_r = (r - 0.5)/n$. A plot of $ud(p)$ against $ld(p)$ for the data points will give a 45° straight line for a perfectly symmetrical distribution of data. The $ld(p)$ and $ud(p)$ can be used to define the **inter-p-range**, sometimes called a quasi-range or spread function, and the **p-difference** by

$$ipr(p) = \tilde{Q}(q) - \tilde{Q}(p) = ld(p) + ud(p)$$

$$pd(p) = ud(p) - ld(p) = \tilde{Q}(q) + \tilde{Q}(p) - 2m.$$

The p-difference will be approximately zero for a symmetrical distribution. To provide a measure of skewness, $pd(p)$ can be standardized by dividing by iqr or $ipr(p)$ to remove the influence of scale. These thus give

The **Galton p-skewness** $g(p) = pd(p)/iqr$,

The **p-skewness index** $g^*(p) = pd(p)/ipr(p)$.

It will be seen that as with the ordinary Galton skewness, $g^*(p)$ lies in $(-1, 1)$. A further quantile-based measure that is sometimes used, which replaces differences by ratios, is the **skewness ratio**:

$$sr(p) = ud(p)/ld(p).$$

For a set of data these measures can be evaluated for the p_r percentage points of the data. Thus for $g^*(p)$, for example, the function of the data that can be plotted as a function of p_r is

$$g^*(p_r) = (x_{(r)} + x_{(n+1-r)} - 2m)/(x_{(n+1-r)} - x_{(r)}).$$

The use of the maximum of the $g(p)$ over $0 < p < 0.5$ to give an overall measure of asymmetry is suggested by MacGillivray (1992).

A further measure of skewness follows the logic of the standard deviation. It is based on the average of the cubes of the deviations from the average, d_i. If there is a long tail to the right, the positive deviations will tend to be larger than the negative deviations so a positive average of the deviations cubed will be obtained. The definition is thus

$$m_3 = \Sigma \, d_i^3 / n \, .$$

The notation m_3 relates to the terminology that such quantities are **moments about the mean**, since the d refers to deviations from the mean. (Notice, for example, that $m_2 = s^2$, but with a devisor of n rather than $(n - 1)$). The skewness thus relates to the third moment about the mean. As with Galton's index it is useful to remove the effect of the magnitude of the numbers, i.e., the scale. An index that achieves this is the **skewness index**:

$$b_1^2 = m_3^2 / m_2^3 \, .$$

It will be seen that this is independent of the scale used.

2.5   Other measures of shape

Thus far we have looked at measures of position, variability and skewness. A further feature of a data set is often referred to as its kurtosis. This refers to the peakiness or flatness of the centre of a distribution, with its consequent link to the lengths of the tails. Peakiness is often associated with long tails and flatness in the centre of the distribution with short tails since, for example, the flat middle of the distribution draws in probability from the tails. These aspects of the shape of a distribution can be measured in a number of ways.

One feature that we have mentioned is the inter p-range. If this is standardized relative to the interquartile range then we have a function, $t(p)$, $0 \leq p \leq 0.5$, that gives an indication of the extension of the tails of the distribution of the data for small p, and of the central concentration for p close to 0.5. Thus the **shape index** is given by

$$t(p) = ipr(p)/iqr.$$

It is evident from the definition that $t(0.25) = 1$ and $t(0.5) = 0$. For p close to 0.5, a peaky distribution will tend to have a smaller $t(p)$ than a less peaky one. Conversely, for small p out in the tails, a long-tailed distribution will have large $ipr(p)$ and relatively large $t(p)$. This measure is particularly suited to symmetric distributions. For non-symmetric distributions it is useful to look at the tails separately using, for example,

$$ut(p) = ud(p)/uqd \text{ and } lt(p) = ld(p)/lqd.$$

If one wishes to compare data with specific distributions it is always possible to create relative measures out of the measures we have been discussing. For example, suppose we wish to compare the data with the commonly used normal distribution, whose quantile function $N(p)$ is readily available. We calculate $T_N(p)$ using normal population values and form the ratio $t(p)/T_N(p)$. If, for example, this is appreciably greater than one for small p, then the data is from a distribution with longer tails than the normal distribution.

A classic term for the flatness/peakiness, and consequent tail shape, of a distribution is **kurtosis**. A simple quantile measure of this makes use of the octiles, $e_j = \tilde{Q}(j/8)$. The Moors kurtosis measure is

$$k = [(e_7 - e_5) + (e_3 - e_1)]/iqr,$$

(see, for example, Moors (1988)). The four values (m, iqr, g, k) provide a simple quantile-based summary of the shape of a distribution.

Another way of looking at kurtosis and tail shape considers the two tails separately. Consider the right tail on its own as a distribution. The relative weight of the peak and tails will be measured by its skewness. If we use the p-difference to measure this, remembering that we actually have only the right-hand half of the data, centred in fact at the upper quartile, we obtain an **upper kurtosis** measure:

$$uk(t) = [\tilde{Q}(1 - t) + \tilde{Q}(0.5 + t) - 2\tilde{Q}(0.75)]/[\tilde{Q}(1 - t) - \tilde{Q}(t)], \ 0 < t < 0.25$$

with a corresponding $lk(t)$ for the lower half of the data. (For data that show symmetry, the data could be replaced by the absolute values $ABS(x - m)$ and the usual qd applied.)

The measurement of both skewness and kurtosis was first investigated by Sir Karl Pearson between 1890 and 1910. His commonly used measure of kurtosis is based on sample moments and uses the fourth moment about the mean:

$$m_4 = \Sigma \, d_i^4/n.$$

A standardized **index of kurtosis** is given by

$$b_2 = m_4/m_2^2.$$

If the data are spread out in a fairly uniform fashion, i.e., the distribution of the data is fairly flat, the value of b_2 will be small. If there is a high concentration in the centre with a few outlying observations, the distribution is regarded as peaky and b_2 will be large. The normal distribution has a population value of the index of three. The index b_1 is more sensitive than s to the effects of outlying observations and b_2 is even less robust. The skewness and kurtosis indices are available as standard in most relevant software. It has been observed that there are variations in the calculation of skewness and kurtosis coefficients between different computer packages (see Joanes and Gill (1998)). Sometimes $b_2 - 3$ is used in place of b_2 and sometimes "corrections" are introduced to get "better" values for small samples. It should also be noted that b_2 depends on both central and tail data and that very different shaped data can lead to the same b_2. The average, standard deviation, and b_1 and b_2 provide a moment-based summary of a set of data. They have been used since the early days of this century to both describe and identify distributions. They do not, however, provide a unique identification. As we have seen, there are many quantile-based quantities that produce parallel but generally more robust measures of the shape of data sets. We, therefore, concentrate on using the raw data with supplementary quantile-shape measurements as the basis for our study.

2.6   Bibliographic notes

In a paper of 1875 Francis Galton presented a plot of the ordered heights, $x_{(r)}$, of a large number of men against the proportion of the sample up to each height $p = r/n$. He marked on the plot the values for $p = 1/4$, $1/2$ and $3/4$ as being of special interest. In a book a few years later (1883) he shows a different plot but with the same axes and terms it the "ogive." Both pictures are reproduced in *The History of Statistics* by Stigler (1986). I was always taught that the ogive was the CDF, $F(x)$, but Galton's plots are clearly of the quantile function, both empirical and in relation to $N(p)$ for the normal distribution. Indeed he uses it to define medians and quantiles which were formally introduced in 1882. Galton clearly saw the importance for data analysis of ordering the data. In the years following, the mathematical foundations of statistics were in a state of rapid development, particularly through the work of Karl Pearson (e.g., Pearson (1895)). This involved great interest in the normal distribution and its definition via the PDF, of necessity as this is its only explicit form, and distributions generally were studied via their PDF and CDF. It would seem that Galton's use of ordered data and the quantile function were almost forgotten and the term quantile only appears to have been first introduced by M. G. Kendall in 1940. There was a continued interest in the properties and specific applications of order statistics, but they were studied via their CDF and PDF. The first books on order statistics did not appear until the second half of the 20th century, e.g., a research book Sarhan and Greenberg (1962) and the first textbook, David (1970). For a further general history, see Hald (1998).

The use of quantiles in data analysis was re-emphasized a hundred years after Galton, particularly by Tukey, e.g., Tukey (1970). Emanuel Parzen, beginning with a classic paper in 1979, introduced a wide range of quantile-based methods and, in particular, the use of the quantile function as an important way of defining a distribution. He has been the main exponent of the use of quantile-based methods, often, but not exclusively, in non-parametric studies of data. Particular papers of relevance are Parzen (1979), (1993) and (1997).

There has recently been much interest in issues of quantile-based measures of shape and kurtosis, but no universally agreed outcomes. Some illustrative papers are those of Balanda and MacGillivray (1988), Groeneveld (1998), Ruppert (1987), and Brooker and Ticknor (1998).

The plotting of data via quantile plots goes back at least to Hazen (1914). The use of p-frequency density plots is much more recent. Parzen (1979) introduced the concept and Jones and Daly (1995) illustrate the use of the sample p-PDF as the basis for exploring data.

At the end of the 19th century Pearson developed a family of four-parameter distributions, the Pearson distributions, that included several now common distributions. He showed that these could be uniquely characterized using the four measures of position, scale, skewness, and kurtosis, estimated for a sample by ($\bar{x}$, s^2, b_1, b_2). These four measures have been widely used ever since as a general means of studying the distributions of samples. However, as they only measure four specific features, in a rather non-robust manner, they do not give a universal practical or unique means of matching sample and distribution. A parallel and more robust approach using quantile measures is discussed in Moors et al. (1996).

There is a large literature on the empirical formulae best used for calculating quantiles for any p and in particular on formulae that lead to smooth curves. We do not go into the details of these methods in the context of this introductory text on parametric modelling. The literature does, however, provide for far more sophisticated ways of carrying out what we see here as the smoothing of plots. Relevant references are

Sheather and Marron (1990), Dielman, Lowry and Pfaffenberger (1994), Cheng and Parzen (1997), and Hutson (in press). It should be noted that the main reason for not treating this subject is that when we have an assumed model our concern is mainly to find only those $p_{(r)}$ corresponding to the ordered observations $x_{(r)}$. It should also be noted that Parzen, e.g., Parzen (1997), has shown how to unify many quantile-based ideas to cover data for discrete variables as well as for continuous variables.

2.7   Problems

1. The following data gives the diameters of a set of 20 metal rods produced by a casting process. The measurements are all of the form 19.-- mm, the table giving the final two figures. Derive the five-number summary: the median, range and interquartile range, the skewness and kurtosis index, the Galton skewness coefficient and $t(0.9)$.

    41, 50, 31, 44, 47, 56, 60, 66, 42, 38, 35, 27, 39, 42, 52, 51, 49, 51, 47, 45

2. A firm sells metal coils of varying weights (Tn) to order. In a given week 91 coils are sold with weights as in the following frequency table. Calculate the main quantile-based measures and plot the quantile function as estimated by plotting the upper class intervals against the p calculated from the frequencies. Obtain the differences Dx and Dp and hence a plot of the empirical p-PDF.

class interval	frequency
10^+–12	2
12^+–14	3
14^+–16	11
16^+–18	18
18^+–20	35
20^+–22	20
22^+–24	2
Total	91

3. Explore further some data you are very familiar with using the measures discussed in this chapter.

CHAPTER 3

Describing a Population

3.1 Defining the population

The word statistics has its origin in the numerical study of states and their populations. How big is the population? How many are there to pay the taxes or join the army? As a consequence of this original usage we use the term **population** to represent all the possible observations that might have been made. For the case of a state where we do, say, a one in ten sample as part of a full census, the sample comes from the finite population, which is the population of the state. If we take a sample of 50 items manufactured on a machine as part of a quality inspection, the sample is envisaged as coming from the theoretically infinite population of items that could have been produced on that machine. The bigger the sample taken the more likely it is to reflect the statistical characteristics of the population. We now need to define more formally the population and these statistical characteristics. That is the purpose of this chapter.

In Chapter 1 we described a population as a whole by the use of five different but interrelated functions:

The Cumulative Distribution Function	$F(x)$ $[= Pr(X \leq x)]$
The Probability Density Function	$f(x)$ $[= dF(x)/dx]$
The Quantile Function	$Q(p)$ $[= F^{-1}(p)]$
The Quantile Density Function	$q(p)$ $[= dQ(p)/dp]$
The p-Probability Density Function	$f_p(p)$ $[= f(Q(p)) = 1/q(p)]$

Our definitions kept to the case where the random variable is continuous, and hence $F(x)$ and $Q(p)$ are continuous increasing functions. We have seen that the inversions or integrations required to move from one form to another may not always be analytically possible, so not all these distributional forms may have explicit formulae.

Distributions may be defined in any of the forms, so it may be necessary to use numerical approximations to find the values of the non-explicit functions.

Using the quantile function as the basis for describing a population led us in Chapter 1 to describe particular features of the population by substituting values for p, such as 0.5, to give the median. Thus we can describe the overall functional shape of a population and also highlight specific features. In this chapter, as in this book generally, we will concentrate on the quantile approach. However, we must also remember that this is a supplement to the classical approaches and not a total alternative. We need, therefore, to keep an eye on the links between, for example, the quantile-based measures and those related to the CDF and PDF.

3.2   Rules for distributional model building

A theme of this book, introduced in Chapter 1, is the value of quantile functions and quantile-based approaches in the building of empirical models for sets of data. As with construction toys we build sophisticated models out of simple components. We have illustrated this construction with some examples. In this section we formalize the methodology by stating a set of construction rules that quantile functions obey. Although we are concentrating on continuous distributions, these rules also apply to the distributions of discrete variables, for which the condition that $Q(p)$ is continuous increasing becomes the requirement that it is non-decreasing, i.e., it may contain step increases and horizontal sections. These conditions are assumed throughout.

The reflection rule

The distribution $-Q(1 - p)$ is the reflection of the distribution $Q(p)$ in the line $x = 0$.

To see this, just consider some illustrative values:

p	$Q(p)$	$-Q(1 - p)$
0	$Q(0)$	$-Q(1)$
0.25	LQ	$-UQ$
0.5	M	$-M$
0.75	UQ	$-LQ$
1	$Q(1)$	$-Q(0)$

If the distribution lies in $(0, \infty)$, the reversed distribution will lie in $(-\infty, 0)$. In general then the distribution $-Q(1 - p)$ is the distribution of the random variable $-X$.

Example 3.1: The quantile function $[-\ln(1-p)]^{\beta}$ is a Weibull distribution with a distributional range $(0, \infty)$. The quantile function $Q(p) = [\ln(p)]^{\beta}$ is a distribution in the range $(-\infty, 0)$ that looks like the reflection of the Weibull in a mirror set at $x = 0$. This distribution is called the reflected Weibull.

The addition rule

If $Q_1(p)$ and $Q_2(p)$ are quantile functions, then $Q_1(p) + Q_2(p)$ is also a quantile function.

This follows from the simple fact that all we require of a quantile function is that it is a non-decreasing function of p in $0 \leq p \leq 1$. Clearly the sum of two such functions must itself be non-decreasing. We illustrated this rule in Example 1.15 and Figure 1.8.

The multiplication rule for positive variables

For distributions where $Q(p)$ can take on negative values, the product of two quantile functions is not necessarily non-decreasing. However, statisticians are often interested in areas of statistics, such as reliability, where the variables are inherently positive. In such cases the product of two quantile functions is inherently non-decreasing, so we have the rule:

The product of two positive quantile functions, $Q(p) = Q_1(p) \times Q_2(p)$, is also a quantile function.

This rule was illustrated in Example 1.19.

The intermediate rule

If $Q_1(p)$ and $Q_2(p)$ are two quantile functions, then the QF

$$Q_i(p) = w.Q_1(p) + (1 - w).Q_2(p), \ 0 \leq w \leq 1,$$

lies between the two distributions.

That is to say that if for a given p, $Q_1(p) < Q_2(p)$ then

$$Q_1(p) \le Q_i(p) \le Q_2(p).$$

Example 3.2: Using the exponential and reflected exponential we can define a distribution

$$Q(p) = 0.2 \ln p - 0.8 \ln (1 - p).$$

The quantile function of this distribution will lie between those of the exponential and the reflected exponential, the values of the weights determining that it will be closer to the exponential over most of the distributional range. This as we have seen is the skew logistic distribution.

The standardization rule

If a quantile function S(p), has a standard distribution in the sense that some measure of position, e.g., the median, is zero and some linear measure of variability, scale, e.g., the IQR, is one, then the quantile function

$$Q(p) = \lambda + \eta\, S(p),$$

has a corresponding position parameter of λ and a scale parameter of η.

Thus the two parameters λ and η control the position and spread of the quantile function. Obviously if we know the parameters of a quantile function, we can use this result in reverse to create a standard distribution:

$$S(p) = [Q(p) - \lambda\,]/\eta,$$

Example 3.3: Consider the symmetric logistic distribution again. Substituting $p = 0.25, 0.5, 0.75$ gives

$$LQ = -\ln(3)$$

$$M = 0$$

$$UQ = \ln(3)$$

so

$$IQR = 2 \ln(3).$$

It follows that the distribution

$$Q(p) = \ln(p/(1 - p))/(2 \ln(3))$$

has zero median and an *IQR* of one and thus acts as a **standard distribution**. From this it follows that a general logistic distribution with median λ and *IQR* η is given by

$$x_p = \lambda + \eta \ln(p/(1 - p))/(2 \ln(3)).$$

Often it is convenient to use not the standard distribution but the form of the model with the simplest mathematical form, here $\ln(p/(1 - p))$. These simple forms we have termed the basic distributions.

The reciprocal rule

The distribution of the reciprocal of a variable, $1/X$, cannot simply be obtained from $1/Q(p)$, since this is a decreasing function of p. However, as with the reversal of x, the roles of p and $1 - p$ are interchanged in the reciprocal, but the sign remains the same. Thus,

the quantile function for the variable $1/X$ is $1/Q(1 - p)$.

To see this denote $1/x$ by y and let the p-quantile of y be y_p with corresponding value x_p. By definition

$$p = \text{Prob}(Y \leq y_p) = \text{Prob}(1/X \leq 1/x_p) = \text{Prob}(x_p \leq X) = 1 - \text{Prob}(X \leq x_p)$$

Hence $x_p = Q(1 - p)$ and therefore $y_p = 1/Q(1 - p)$.

Example 3.4: The power distribution has *QF*, p^β, where $\beta > 0$. The distribution of $y = 1/x$ for this distribution is, by the rule: $Q_1(p) = 1/(1 - p)^\beta$, which is the Pareto distribution.

The Q-transformation rule

If $z = T(x)$ is a non-decreasing function of x, then $T(Q(p))$ is a quantile function.

This rule holds since a non-decreasing function of a non-decreasing function must itself be non-decreasing. Conversely, if $T(x)$ is a non-increasing function of x, then $T(Q(1 - p))$ is a *QF*. We will term such transformations of the quantile function **Q-transformations**. It will be shown in Chapter 4 that these derived quantile functions are in fact the quantile functions of the transformed variable Z.

Example 3.5: Consider $Q(p)$ to be an exponential and the transformation be $T(x) = x^\beta$, where β is a positive power. We then have a Q-transformation as

$$Q_t(p) = [-\ln(1 - p)]^\beta.$$

This is a quantile function and is in fact the quantile function for a much used distribution that we have already met called the Weibull distribution.

The uniform transformation rule

If U has a uniform distribution then the variable X, where $x = Q(u)$, has a distribution with quantile function $Q(p)$. Thus data and distributions can be visualized as generated from the uniform distribution by the transformation $Q(.)$, where $Q(p)$ is the quantile function.

  This rule follows directly from the Q-transformation rule when it is observed that the quantile function of the uniform distribution is just p.

The p-transformation rule

If $H(p)$ is a non-decreasing function of p in the range $0 \leq p \leq 1$, standardized so that $H(0) = 0$ and $H(1) = 1$, then $Q(H(p))$ is also a quantile function, with the same distributional range as $Q(p)$.

  We will refer to transformations of p within a quantile function as p-transformations. We have seen numerous plots of quantile functions, i.e., plots of $Q(p)$ against p. A Q-transformation transforms the vertical $Q(p)$-axis. The p-transformation transforms the p-axis. In both cases the conditions ensure that the resulting plot is still of a valid quantile function.

Example 3.6: Noting that the positive power transformation of p, p^α, satisfies the conditions, we can use the p-transformation on the Weibull distribution, referred to above, to give

$$x_p = [-\ln(1 - p^\alpha)]^\beta, \text{ where } \alpha, \beta > 0.$$

This again is a named distribution, the Generalized Exponentiated Weibull.

3.3   Density functions

Three density functions were introduced in Chapter 1: the probability density, $f(x)$; the quantile density, $q(p)$; and the p-probability density, $f_p(p)$. The terminology refers to the concentration, density, of (a) the probability along the x-axis for $f(x)$; (b) the quantiles along the probability axis for $q(p)$; and (c) the probability along the x-axis expressed in terms of the total probability up to that point for $f_p(p)$. If we refer back to the addition rule of the last section and differentiate we obtain:

The addition rule for quantile density functions

The sum of two quantile density functions is itself a quantile density function:

$$q(p) = q_1(p) + q_2(p).$$

To be precise if we obtain a quantile function by adding two quantile functions together its quantile density function is obtained by the addition of the quantile density functions of the component parts. Since $f_p(p) = 1/q(p)$, then the corresponding relation for p-PDF is

$$f_p(p) = 1/[1/f_{1p}(p) + 1/f_{2p}(p)].$$

If we start with a quantile density function, the quantile function has to be obtained by integration. In this case the position parameter λ appears as a constant of integration.

Notice the information on the general shape of a distribution given by substituting $p = 0$ and $p = 1$ in $Q(p)$ and $f_p(p)$. The first gives the range of the X variable, its distributional range, $(Q(0), Q(1))$; the second gives the ordinates of the probability density at the two ends of the distribution.

Example 3.7: For the unit exponential distribution $S(p) = -\ln(1 - p)$, with $f_p(p) = 1 - p$, we get

$$DR = (0, \infty), f_p(0) = 1, f_p(1) = 0.$$

3.4   Population moments

To develop the ideas of the population quantities analogous to the sample moments discussed in Chapter 1, consider a positive variable such as the weights of some metal castings. Suppose one third of the castings weighed 3 kg and two thirds weighed 6 kg. The average weight, the first sample moment, would be

$$3 \times 1/3 + 6 \times 2/3 = 3.$$

Notice that with the information in this form we do not need to know the actual numbers of castings, only the values and the proportions. Thus the sample mean can be expressed as

$$\bar{x} = \Sigma \ (x. \ \text{proportion}).$$

Turning now to the corresponding population situation, Figure 3.1(a) shows the population broken into infinitesimal proportions, dp, for each of which the value of x is given by $x = Q(p)$. The summation becomes an integral, so the **population mean**, μ, is defined in an analogous fashion to the sample mean by

$$\mu = \int_0^1 Q(p)dp \,.$$

This is also referred to as the **expectation, $E(X)$**, of the random variable X. It is the value we expect on average to see. A slightly different way of seeing this result is illustrated in Figure 3.1(b). The area under the quantile function, $Q(p)$, is the integral on the right of the definition, which equals the area under the line at height, μ, since the base has unit length. For the sake of clarity of exposition we have used positive variables in the discussion. Exactly the same definitions and argument apply when X can take negative values. However, in this case areas under the p-axis count as negative; for example, with a distribution that is symmetric about $x = 0$, the positive and negative areas will cancel, giving a mean of zero. To see the relation of the definition in terms of quantile functions to that in terms of probability density functions, we remember that

$$dp = f(x) \ dx \ \text{and} \ x = Q(p),$$

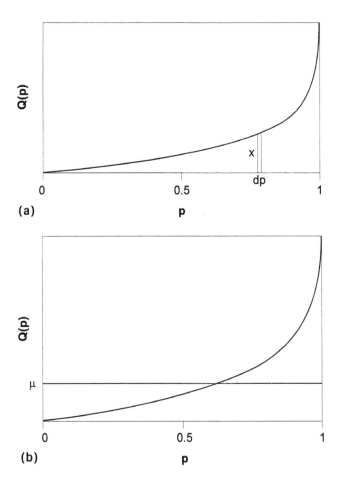

Figure 3.1. (a) Defining moments; (b) The mean

so
$$\mu = \int_{DR} x\,f(x)\,dx\,.$$

The limits of 0 and 1 for p become whatever is the distributional range for x, usually (0 to ∞) or ($-\infty$ to $+\infty$). Consider some examples.

Example 3.8: For the Pareto distribution

$$x_p = 1/(1-p)^\beta,\ \beta > 0.$$

Thus
$$\mu = \int_0^1 (1-p)^{-\beta} dp$$

$$= [-(1-p)^{(1-\beta)}/(1-\beta)]_0^1$$
$$= 1/(1-\beta), \quad \text{provided } 0 < \beta < 1.$$

The fact that, although the distribution is defined for all positive β, the mean only exists for $0 < \beta < 1$ is due to the long positive tail that for β greater than one leads to an infinite mean.

A simple and useful result flows directly from the definition of the expectation. If η and λ are constants, then substituting in the definition gives

$$E(\lambda + \eta X) = \lambda + \eta \ E(X).$$

More generally for a non-decreasing transformation, $Y = T(X)$, the distribution of Y is $T(Q(X))$ by the Q-transformation rule. It follows that

$$E[T(X)] = \int_0^1 T[Q(p)] dp \ .$$

Higher moments are defined to provide measures of spread, skewness and kurtosis. Table 3.1 gives a summary. The higher moments about the mean can be obtained by expansion from the central moments, which are more directly calculated (Table 3.2 illustrates).

Example 3.9: Returning to the Pareto distribution for illustration we have

$$E(x^2) = \int_0^1 [1/(1-p)^{2\beta}] dp$$

By analogy with the previous calculation, this is seen to be

$$\mu_2' = 1/(1 - 2\beta), \text{ for } 0 < \beta < 0.5.$$

Notice that we now have an even tighter constraint to give a finite variance. On substituting in the expression for μ_2 and simplifying, we finally obtain

$$\mu_2 = \beta^2/[(1 - 2\beta)(1 - \beta)^2], \text{ for } 0 < \beta < 0.5.$$

Moment / Quantity	Symbols	Definition
Second central moment	μ_2'	$E(X^2) = \int_0^1 [Q(p)]^2 dp$
Variance (second moment about the mean)	$\mu_2 = V(x)$	$E[(X-\mu)^2] = \int_0^1 [Q(p)-\mu]^2 dp$
Standard deviation	σ	$\sigma = \sqrt{\mu_2}$
Third moment about the mean	μ_3	$E[(X-\mu)^3] = \int_0^1 [Q(p)-\mu]^3 dp$
Index of skewness	β_1^2	μ_3^2/μ_2^3
Fourth moment about the mean	μ_4	$E[(X-\mu)^4] = \int_0^1 [Q(p)-\mu]^4 dp$
Index of kurtosis	β_2	μ_4/μ_2^2
Linear transformation		$V(aX+b) = a^2 V(X)$
General central moments	μ_r'	$E(X^r) = \int_0^1 [Q(p)]^r dp$
General moments about the mean	μ_r	$E[(X-\mu)^r] = \int_0^1 [Q(p)-\mu]^r dp$

Table 3.1. Definitions of population higher moments

$$\mu_1 = \mu_1'$$
$$\mu_2 = \mu_2' - \mu_1^2$$
$$\mu_3 = \mu_3' - 3\mu_2'\mu_1 + 2\mu_1^3$$
$$\mu_4 = \mu_4' - 4\mu_3'\mu_1 + 6\mu_2'\mu_1^2 - 3\mu_1^4$$

Table 3.2. Relations between moments about the mean and central moments

3.5   Quantile measures of distributional form

In Chapter 1 and primarily in Chapter 2, a range of measures of the form of distributions were introduced based on the sample quantiles. Chapter 1 introduced a few population analogues for the median, interquartile range, etc. In this section the list of population quantile measures of shape is further extended. However, as the new ones introduced are analogues of sample measures already discussed in Chapter 2, the list of measures and their definitions is condensed to Table 3.3.

Name	Symbol	Definition
Median	M	$Q(0.5)$
Lower quartile	LQ	$Q(0.25)$
Upper quartile	UQ	$Q(0.75)$
Interquartile range	IQR	$UQ - LQ$
Inter p-range	$IPR(p)$	$Q(1 - p) - Q(p)$
Shape index	$T(p)$	$IPR(p)/IQR$
Quartile difference	QD	$LQ + UQ - 2M$
p-Difference	$PD(p)$	$Q(p) + Q(1 - p) - 2M$
Galton's skewness	G	QD/IQR
Galton's p-skewness	$G(p)$	$PD(p)/IQR$
p-Skewness index	$G^*(p)$	$PD(p)/IPR(p)$
Upper p-difference	$UD(p)$	$Q(1 - p) - M$
Lower p-difference	$LD(p)$	$M - Q(p)$
Upper shape index	$UT(p)$	$UD(p)/(UQ - M)$
Lower shape index	$LT(p)$	$LD(p)/(M - LQ)$
Skewness ratio	$SR(p)$	$UD(p)/LD(p)$
Moors' kurtosis	K	$[(Q(7/8) - Q(5/8))+ ((Q(3/8) - Q(1/8))]/IQR$
Upper kurtosis	$UK(t)$	$Q(1 - t) + Q(0.5 + t) - 2Q(0.75)$
Upper kurtosis index	$UKI(t)$	$UK(t)/[Q(1 - t) - Q(0.5 + t)]$
Lower kurtosis	$LK(t)$	$Q(t) + Q(0.5 - t) - 2Q(0.25)$
Lower kurtosis index	$LKI(t)$	$LK(t)/[Q(0.5 - t) - Q(t)]$

Note: $0 \le p \le 0.5, 0 \le t \le 0.25$.

Table 3.3. Definitions of quantile measures of shape

Notice that all definitions that involve addition or subtraction are independent of any position parameter and all those involving a ratio are independent of both position and scale parameters. We now illustrate the definitions further by some examples.

Example 3.10: As the simplest possible example consider the uniform distribution which has $S(p) = p$. Thus $LQ = 0.25$, $UQ = 0.75$ and $M = 0.5$, hence $IQR = 0.5$. The p-quantiles are $1 - p$ and p. The distribution is thus symmetrical about 0.5 and G and $G(p)$ are both zero. For the shape index we have $T(p) = 2(1 - 2p)$, so, for example, $T(0.05) = 1.8$, the high value indicating the flatness of the distribution.

Example 3.11: To illustrate further consider a distribution that is formed by the weighted sum of a basic distribution $S(p)$, which may or may not contain further shape parameters, and its reflection, $-S(1 - p)$. A reasonable practical requirement is that $S(p)$ has distributional range $(0, a)$ where the positive a may be ∞. We illustrated this idea with the skew logistic in Chapter 1. A convenient way to introduce the general skewness parameter into the weighting is to define the distribution by

$$Q(p) = \lambda + (\eta/2) \ [(1 + \delta) \ S(p) - (1 - \delta) \ S(1 - p)], \ -1 \leq \delta \leq 1.$$

The distribution lies in the range of $(\lambda - a \ (\eta/2) \ (1 - \delta), \ \lambda + a \ (\eta/2) \ (1 + \delta))$, which for $a = \infty$ is $(-\infty, \infty)$. The quantile density, using $s(p)$ for the quantile density of $S(p)$, is

$$q(p) = (\eta/2)[(1 + \delta) \ s(p) + (1 - \delta) \ s(1 - p)].$$

A study of the quantile statistics of this distribution will show why this form has been adopted. Notice that using R to denote the reflected distribution we have that $M_R = -M_S$, $UQ_R = -LQ_S$ and $LQ_R = -UQ_S$. Substituting $p = 0.5$ gives

$$M = \lambda + (\eta/2) \ [(1 + \delta) \ M_S + (1 - \delta) \ M_R]$$

$$= \lambda + \eta\delta \ M_S.$$

From the quartiles we obtain

$$\begin{aligned}
IQR &= \{\lambda + (\eta/2)[(1 + \delta)UQ_S + (1 - \delta)(-LQ_S)]\} \\
&\quad -\{\lambda + (\eta/2)[(1 + \delta)LQ_S + (1 - \delta)(-UQ_S)]\} \\
&= \eta IQR_S
\end{aligned}$$

Similarly, we obtain $IPR(p) = \eta \ IPR_S(p)$ and hence for the shape index we have $T(p) = T_S(p)$. Thus the shape index of $Q(p)$ is the same as that of $S(p)$. Note that it does not depend on the position, scale, or skewness parameters of the model, but only on any shape parameter(s) in $S(p)$. This result is the basis for the use of the term shape index for this measure and is a consequence of the choice of weights that add to one for all δ. Turning to the p-difference function, we obtain

$$\begin{aligned}
PD(p) &= (\eta/2) \ [(1 + \delta)S(p) - (1 - \delta)S(1 - p)] \\
&\quad + (\eta/2)[(1 + \delta)S(1 - p) - (1 - \delta)S(p)] - \eta\delta \ M_S,
\end{aligned}$$

which simplifies to

$$PD(p) = \eta\delta \ PD_S(p).$$

For the quartiles, this gives

$$QD = \eta\delta \, QD_S.$$

It follows that the skewness measures are

$$G = \delta G_S, \quad G(p) = \delta G_S(p), \quad \text{and } G^*(p) = \delta G^*{}_S(p).$$

Thus the parameter δ controls the skewness, with positive values giving a positively skewed distribution. It has no effect on the spread as measured by the inter-p ranges but does influence the value of the median. The η acts as a multiplier to the inter-p ranges and is thus a scale parameter. The parameter λ does not influence the ranges or skewness measures but does have the effect of shifting the median. It thus controls the position of the distribution. As we saw with the skew logistic this general form of distribution has three parameters that link directly and independently to the three main features of the shape of the distribution. Any shape parameters link with the p-spread function, $T(p)$.

As all the measures discussed above are based directly on $Q(p)$, it is of some value to note the converse relation of $Q(p)$ to the various p-measures. Consider, and check, the following identity:

$$Q(p) \equiv Q(0.5) + \{Q(p) - Q(1 - p)\}/2 + \{Q(p) + Q(1 - p) - 2Q(0.5)\}/2$$

so

$$Q(p) \equiv M + IQR.T(p) \, (1 + G^*(p))/2.$$

Thus for symmetric distributions $Q(p) = M + IQR.T(p)/2$. For skew distributions, $G^*(p)$ contributes to the expression. Thus $Q(p)$ is seen as made up, using p values, of the contributions from position, scale, p-skewness index and shape index. This suggests that in studying distributional models these four quantities should play a significant role, as they uniquely define the quantile function.

3.6   Linear moments

L-moments

In the previous sections we introduced third and fourth moments to describe the skewness and kurtosis of a distribution. The use of even

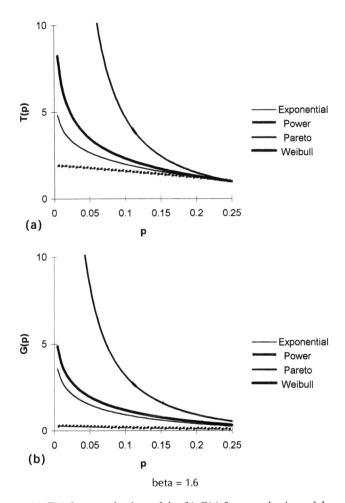

Figure 3.2. (a) $T(p)$ for some basic models; (b) $G(p)$ for some basic models

higher order moments was also possible in the definitions. Unfortunately, as soon as we raise observations to third, fourth and higher powers we obtain statistics that have very large inherent variability. If we handle models with more than two parameters, the statistics based on ordinary moments often requires us to use such imprecise measures. It would be much better if we could obtain moment-like statistics describing the various aspects of the shape of a distribution using just linear functions of the order statistics. The definitions of the inter-p range and p-difference indicate ways in which this may be done. Consider now a measure of spread based on the average of the ranges of all pairs of observations taken from a sample of n, i.e., the average of v_{ij}, where $v_{ij} = x_j - x_i$, $1 \leq i < j \leq n$. There are $\binom{n}{2}$ such pairs, sub-samples. The value $x_{(r)}$ will occur in some with a plus (+) sign, in others with a minus (−) sign. If the value of v_{ij} is summed over all combinations, the coefficient of the r-th order statistic, the sum of all the + and − is $K_r = (2r - n - 1)$. Thus a sample measure of spread given by the average v_{ij} is

$$l_2 = [\Sigma_r \ (2r - n - 1)x_{(r)}]/\binom{n}{2}.$$

Notice that unlike m_2 this is a linear function of the observations. Such moments are called **L-moments**, L for Linear (see, for example, Hosking (1990)). The sample L-moment for skewness is based on the average skewness coefficient based on samples of three, $x_{(k)} - 2x_{(j)} + x_{(i)}$, $1 \leq i < j < k \leq n$. A kurtosis measure using samples of four observations is obtained by using the inner two observations to give the iqr, the outer two to give the ipr, and setting $t(p) = 3$ as a reasonable value for comparison. This leads to the simple form: $x_{(l)} - 3x_{(k)} + 3x_{(j)} - x_{(i)}$, where $1 \leq i < j < k < l \leq n$, as a measure of kurtosis for four observations. The L-moment is then obtained by averaging over all possible quadruples formed from the n observations. The first order sample L-moment is obviously the average over samples of one, so is no different from the ordinary average. The essence of all these quantities is that they provide measures that are linear in the observations. They thus all have the form

$$l_k = K_k \ \Sigma \ k_r \ x_{(r)}.$$

The actual sample calculation of these moments is best done indirectly and will be returned to later. The population equivalent to this form of calculation is based on using the expected values of the order statistics in small samples. Suppose we denote the r-th order statistic in a sample of n by $X_{r:n}$, then in samples of 1, 2, 3 and 4 we define the first four L-moments by

$$\lambda_1 = E(X_{1:1});$$
$$\lambda_2 = [E(X_{2:2}) - E(X_{1:2})]/2;$$
$$\lambda_3 = [E(X_{3:3}) - 2E(X_{2:3}) + E(X_{1:3})]/3;$$
$$\lambda_4 = [E(X_{4:4}) - 3E(X_{3:4}) + 3E(X_{2:4}) - E(X_{1:4})]/4.$$

It will be evident that λ_1 is just the ordinary mean measuring position, λ_2, λ_3, and λ_4 give the population analogies of the sample quantities just discussed.

In the same way that ordinary moments are combined to give scale-free coefficients we can combine the L-moments to give

The L-Coefficient of Variation $\tau_2 = \lambda_2/\lambda_1$;

The L-Coefficient of Skewness $\tau_3 = \lambda_3/\lambda_2$;

The L-Coefficient of Kurtosis $\tau_4 = \lambda_4/\lambda_2$.

One valuable feature of these coefficients is that they have natural scales with

$$-1 < \tau_3 < 1 \quad \text{and} \quad (5\tau_3^2 - 1)/4 < \tau_4 < 1$$

To evaluate the L-moments for a distribution requires some further theory so we discuss this in Section 4.11.

Probability-weighted moments

A second group of moments are the **probability-weighted moments**, **PWM**. The **population PWM** is defined by

$$\bar{\omega}_{t,r,s} = E[Q^t(p)p^r(1-p)^s],$$

where the expectation treats p as a uniform variable. The weights associated with $Q(p)$ thus depend on p and tend to zero at either end of the distribution. Table 3.4 gives formulae for the PWM of some of

$Q(p)$	$\omega_{r,o}$	$\omega_{o,s}$
1	$1/(r + 1)$	$1/(s + 1)$
p^α	$1/(\alpha + r + 1)$	$B(\alpha + 1, s + 1)$
$(1 - p)^\beta$	$B(r + 1, \beta + 1)$	$1/(\beta + s + 1)$
$p^\alpha/(1 - p)^\beta$	$B(\alpha + r + 1, 1 - \beta)$	$B(1 + \alpha, s + 1 - \beta)$
$[-\ln(1 - p)]^\beta$	—	$\Gamma(1 + \beta)/(s + 1)^{1 + \beta}$

Table 3.4. Probability weighted moments for some basic distributions

the more common distributions. The general definition allows for the variable to be raised to a power, which was avoided with the L-moments. To continue this policy we will keep all the calculations to the case where $t = 1$ and so this subscript will be dropped. Thus interest will focus on the PWM defined by

$$\bar{\omega}_{r, 0} = E[Q(p)p^r] = \int_0^1 Q(p)p^r dp$$

and

$$\bar{\omega}_{0, s} = E[Q(p)(1 - p)^s] = \int_0^1 Q(p)(1 - p)^s dp.$$

Example 3.12: The simplest example is the power distribution with known origin at $x = 0$, $Q(p) = \eta p^\beta$. For such a distribution, the PWMs for $s = 0$ are the easiest to derive. Thus

$$\bar{\omega}_{r, 0} = E[\eta p^{\beta + r}] = \int_0^1 \eta p^{\beta + r} dp = \eta / (r + \beta + 1)$$

In carrying out calculations with PWM it should be noted that the various moments are mathematically related. Consider the following illustrations:

$$\bar{\omega}_{0, 2} = E[Q(p)(1 - p)^2] = E[Q(p)] - 2E[Q(p)p] + E[Q(p)p^2]$$
$$= \bar{\omega}_{0, 0} - 2\bar{\omega}_{1, 0} + \bar{\omega}_{2, 0}$$
$$\bar{\omega}_{0, 0} = E[Q(p)\{p + (1 - p)\}] = \bar{\omega}_{1, 0} + \bar{\omega}_{0, 1}$$
$$\bar{\omega}_{0, 0} = E[Q(p)\{p + (1 - p)\}^2] = \bar{\omega}_{2, 0} + 2\bar{\omega}_{1, 1} + \bar{\omega}_{0, 2}$$
$$\bar{\omega}_{2, 0} = E[Q(p)(1 - q)^2] = E[Q(p)] - 2E[Q(p)q] + E[Q(p)q^2]$$
$$= \bar{\omega}_{0, 0} - 2\bar{\omega}_{0, 1} + \bar{\omega}_{0, 2}$$

In general,

$$\bar{\omega}_{r, 0} = \Sigma_{s = 0}^r \binom{r}{s}(-1)^s \bar{\omega}_{0, s} \text{ and } \bar{\omega}_{0, s} = \Sigma_{r = 0}^s \binom{s}{r}(-1)^r \bar{\omega}_{r, 0}.$$

A consequence of such relations is that if we define the order of a PWM by $k = r + s$, then given the PWM of orders less than k, there is only, in effect, one new PWM of order k. Given any one PWM of k order and the lower order PWMs, all the remaining order k PWMs can be derived mathematically. Hence in developing moments the easiest approach is to keep to the moments with either r or s set at zero.

For the case of PWM the **sample PWM** can be straightforwardly defined by replacing the population expectation by a sample average of the equivalent expression:

$$w_{r,s} = (1/n)\Sigma x_{(i)} p_{(i)}^{*r}(1-p_{(i)}^*)^s$$

where the ordered $p_{(i)}^*$ are some suitably chosen probabilities. The natural $p_{(i)}^*$ to choose are $i/(n+1)$, p_i or the median-p_i, defined in Section 4.2. The values $p_i = (i - 0.35)/n$ have been suggested as leading to statistics with good properties. The form of the sample values ensures that the mathematical relations between the population PWM have analogous formulae for the sample PWMs. We will see in Section 4.11 that L-moments and probability-weighted moments are closely related.

3.7   Problems

1. Consider the following distributions in their basic forms:
   (A) The distribution $S(p) = 1/(1 - p)$, $0 \leq p \leq 1$.
       Distribution A (the reciprocal uniform).
   (B) The distribution $S(p) = 1 - (1 - p)^\beta$, $\beta > 0$.
   (C) A distribution is formed by **adding**
       to distribution A its reversed distribution $Q(p) = -1/p$.
       Thus we have

   $$S(p) = 1/(1-p) - 1/p = (2p - 1)/[p(1-p)]. \text{ Distribution C.}$$

   (D) A further distribution is formed by **multiplying** the
       distribution $1/(1 - p)$ by the uniform quantile function,
       which is just p. Thus we have

   $$S(p) = p/(1 - p). \text{ Distribution D.}$$

   (E) Multiplying the exponential quantile function by p
       gives the distribution

   $$S(p) = -p \ln(1 - p). \text{ Distribution E.}$$

(a) Show that Distribution D is simply Distribution A shifted to make the origin the left end of the distributional range.

Using Distributions B through E defined above, carry out the following exercises:

(b) Find the distributional ranges $[S(0), S(1)]$.

(c) Derive $s(p)$, $f_p(p)$ and hence $f_p(0)$ and $f_p(1)$. Sketch the distributions. Find $f(x)$ and $F(x)$, where possible.

(d) Find the population median, M, interquartile range, IQR, $IPR(p)$, $T(p)$, $D(p)$ and $G(p)$.

(e) Find $E(X)$.

2. Plot the functions $T(p)$, $G(p)$ and $G^*(p)$ for the following distributions:

(a) The exponential distribution.

(b) The Weibull distribution.

(c) The Pareto distribution.

(d) The normal distribution.

(e) The logistic distribution.

3. Consider the Pareto distribution in the form $S(p) = 1/(1-p)^{1/k}$, $k = 1, 2, ...$ Show that the moments μ_r' only exist if $r < k$ and that for these cases they take the values $\mu_r' = k/(k-r)$.

4. (a) If $R(p)$ is the reflected form of $Q(p)$ show that the probability-weighted moments have the following relationships $\omega_{R:\, r,o} = -\omega_{Q:\, o,r}$ and $\omega_{Q:\, o,s} = -\omega_{Q:\, s,o}$.

(b) If a distribution has the form $Q(p) = \lambda + \eta T(p) + \phi H(p)$, where $T(p)$ and $H(p)$ are quantile functions, show that

$$\omega_{Q:\, r,o} = \lambda/(r+1) + \eta\, \omega_{T:\, r,o} + \phi\, \omega_{H:\, r,o},$$

and

$$\omega_{Q:\, o,s} = \lambda/(s+1) + \eta\, \omega_{T:\, o,s} + \phi\, \omega_{H:\, o,s},$$

(c) Using the results of (a) and (b), find the probability-weighted moments, $\omega_{r,o}$ and $\omega_{o,s}$ for the distribution.

$$Q(p) = \lambda + (\eta/2)\,[(1+\delta)p^\alpha - (1-\delta)(1-p)^\alpha].\ -1 \le \delta \le 1.$$

5. Explore the shapes of the following three distributions:
    (a) $Q(p) = 4p^2 - 3p + 1$;
    (b) $Q(p) = 3p^2 - 2p^3$;
    (c) $Q(p) = p^\alpha/(1 - p^\gamma)^\beta$.

CHAPTER 4

Statistical Foundations

4.1   The process of statistical modelling

In Chapter 1 the main elements of the process of statistical modelling were introduced. These were summarized in Section 1.14. The iterative nature of the process should be emphasized. There is no guaranteed way of simply working through steps 1 to 5 to get a good fitted model. Let us illustrate some of the situations where one might need to follow one of the backward loops.

Example 4.1: At the identification stage a trial model shows all the required features, except for one. We then need to go back to the construction stage to see what changes in construction would add the required feature.

Example 4.2: The validation process with a new set of data may indicate a shift in position, but no change in the scale or shape features. Thus a new estimate of the position parameter will be required. The issue is also raised as to whether the combined data, new plus old, might be used to obtain better estimates of the unchanged parameters.

Example 4.3: A model, fitted by, say, the method of percentiles using the median and quartiles, is applied in a new situation. It is realized that for this situation the quality of fit matters more for large values of the variable than for small values. This suggests a need to re-estimate the parameters with a more appropriate choice of quantiles for the matching.

It is evident from these examples that modelling cannot be reduced to a routine process. It is necessary to be constantly on the lookout for features of the data and of the area of application that might lead to a revision, an iteration, of part of the process.

One other general point needs to be made at this stage. Whatever models we finally obtain will be forever tentative. We must not treat them as the truth about the situation. It may be suggested that the classic idea of falsifiability should be used as a final stage in the modelling. We should test our model with some formal test and only use it if it is not rejected by the test. Unfortunately, in most practical situations the real world is a very complex place and our model is inevitably a simplification. Thus, given sufficient data, and in our computer-based world we often have vast quantities, we will reject almost any model we have obtained. The question thus becomes not, is the model right or wrong, or even is it statistically acceptable or not, but, is it good enough to achieve its practical objectives? This is why it is important to make sure the criteria used in processes such as estimation take into account the final practical uses of the model.

In the introduction to this book we emphasized the role of modelling within problem solving and the importance of having new perspectives within that process. In this chapter we introduce a number of common statistical topics and look at them from the viewpoint of quantile modelling. The various topics are introduced not only as the basis for further chapters but also to give a quantile-based perspective on common statistical ideas.

4.2   Order statistics

We previously denoted a set of ordered data by $x_{(1)}, x_{(2)}, ..., x_{(n-1)}, x_{(n)}$, the corresponding random variables being denoted by $X_{(1)}, X_{(2)}, ...,$ $X_{(n-1)}, X_{(n)}$. Thus $X_{(n)}$, for example, is the random variable representing the largest observation of a sample of n. The n random variables are referred to as the n **order statistics**. As we have seen, sample-ordered values play a major role in modelling with quantile-defined distributions, so this section will study some of their main properties. Recent books on this topic are Reiss (1989), Arnold, Balakrishnan and Nagaraja (1992), and the two volumes, 16 and 17, of the *Handbook of Statistics*, edited by Balakrishnan and Rao (1998).

Consider first the distribution of the largest observation. Let $Q(p)$ be the quantile function of the x data and the distribution of $X_{(n)}$ be denoted by $Q_{(n)}(p_{(n)})$. Let the corresponding two CDFs be $F(x)$ $(= p)$ and $F_{(n)}(x)(= p_{(n)})$. The probability that $X_{(n)}$ is less than or equal to some specified value x is given directly as $p_{(n)}$, but is also the probability that all n independent observations on X are less than or equal to this value, x, which for each one is p. This probability is, by the multiplication law of probability, p^n. Hence

$$P_{(n)} = p^n \text{ so } p = p_{(n)}^{1/n} \text{ and } F(x) = p_{(n)}^{1/n}.$$

Finally, therefore, inverting the CDF, $F(x)$, to get the quantile function, we have the equivalent statements that the $p_{(n)}$ quantile of the largest value is given by both $Q_{(n)}(p_{(n)})$ and $Q(p_{(n)}^{1/n})$. Hence

$$Q_{(n)}(p_{(n)}) = Q(p_{(n)}^{1/n}).$$

The quantile function of the largest observation is thus found from the original quantile function in the simplest of calculations.

Example 4.4: If X has an exponential distribution, then the quantile function of the largest observation, returning to the usual notation, is

$$Q_{(n)}(p) = -\ln (1 - p^{1/n}).$$

This distribution will have a median value of $-\ln (1 - 1/2^{1/n})$.

The previous argument also works directly for the smallest observation, using the fact that the probability of n independent observations being greater than or equal to x is $(1 - p)^n$. Thus

$$1 - p_{(1)} = (1 - p)^n \text{ and so } p = 1 - (1 - p_{(1)})^{1/n}$$

This leads to a quantile function for $X_{(1)}$ of $Q(1 - (1 - p_{(1)})^{1/n})$. For the exponential example we have on simplifying,

$$Q_{(1)}(p) = -[\ln (1 - p)]/n.$$

This is still an exponential distribution but with a reduced scale factor.
   For the general r-th order statistic $X_{(r)}$ the calculation becomes more difficult. The probability that the r-th largest observation is less than some value z, $X_{(r)} \leq z$, is equal to $p_{(r)} = F_{(r)}(z)$. In terms of the X variables, this is also the probability that <u>at least</u> r of the n independent observations are less than or equal to z. The probability of s observations being less than or equal to z is p^s, where $p = F(z)$, and there are $n - s$ observations, with probability $(1 - p)^{(n-s)}$ of being greater

than z. There is also the binomial term, $\binom{n}{s}$, giving the number of ways that the s can be chosen from the n. Thus, as Figure 4.1 illustrates, the total probability for s observations less than or equal to z is given by the binomial expression.

$$\text{Prob}(s \text{ observations} \leq z) = \binom{n}{s} p^s (1-p)^{(n-s)}.$$

It follows that the probability of at least r observations less than or equal to z is

$$p_{(r)} = \Sigma_{s=r}^n \text{ Prob}(s \text{ observations} \leq z),$$

since if for any s, for $s = r$ to n, the observation $x_{(s)}$ is less than or equal to z, then $x_{(r)}$ must be less than or equal to z. Hence

$$p_{(r)} = \Sigma_{s=r}^n \binom{n}{s} p^s (1-p)^{(n-s)}.$$

This function is the incomplete beta function and is denoted by

$$p_{(r)} = I(p, r, n-r+1).$$

Clearly, we cannot solve this expression in a simple fashion to get p on the left-hand side. Mathematically this inversion can be carried out numerically and the result referred to as the inverse beta function. We denote it by

$$p = \text{BETAINV}(p_{(r)}, r, n-r+1).$$

We thus have, as with the previous calculations, the relation between the CDF for the order statistic and the CDF for the original distribution of X. Again we can express the value z as either $Q_{(r)}(p_{(r)})$ or $Q(p)$. Hence we have a further modelling rule.

The order statistics distribution rule

If a sample of n observations from a distribution with quantile function Q(p) are ordered, then the quantile function of the distribution of the r-th order statistic is given by

$$Q_{(r)}(p_{(r)}) = Q(\text{BETAINV}(p_{(r)}, r, n-r+1)).$$

Although this looks unhelpful it in fact is a crucial result since BETAINV is a standard function in most spreadsheets and statistical software.

The previous expression for $p_{(r)}$ can be re-expressed in terms of CDF as

$$p_{(r)} = F_{(r)}(z) = \Sigma_r^n \, \binom{n}{s} \, F(z)^s \, [1 - F(z)]^{(n-s)}.$$

To find the PDF of $X_{(r)}$ this is differentiated with respect to z and after some algebra the expression simplifies to

$$f_{(r)}(z) = \{n!/[(r-1)! \, (n-r)!]\} \, [F(z)]^{(r-1)} \, [1 - F(z)]^{(n-r)} \, f(z).$$

Rather than go through the detail of this it is worth noting, looking at the diagram of Figure 4.1, that $f_{(r)}(z)dz$ naturally gives the probability of getting one value of z in the small interval dz with $r-1$ below it and $n-r$ above it. The p-PDF is even clearer as

$$f_{(r)p}(p) = \{n!/[(r-1)! \, (n-r)!]\} \, p^{r-1}(1-p)^{n-r} \, f(p).$$

If we wish to calculate the k-th central moment of the r-th observation of a sample of n, denoted $E(X_{r:n}^k)$ to emphasize the dependence on both r and n, we start with the definition in terms of the PDF $f_{(r)}(z)$, thus

$$E(X_{r:n}^k) = \int_{DR} z^k f_{(r)}(z)dz = \{n!/[(r-1)!(n-r)!]\}$$
$$\int_{DR} z^k [F(z)]^{(r-1)}[1 - F(z)]^{(n-r)} f(z)dz$$

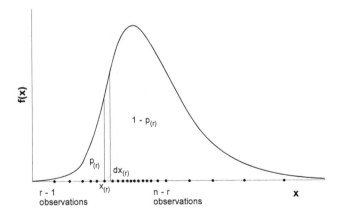

Figure 4.1. Deriving the distribution of the r-th order statistic

changing the variable to $p = F(z)$ leads to

$$E(X_{r;n}^k) = \{n!/[(r-1)!(n-r)!]\}\int_0^1 Q^k(p)p^{(r-1)}[1-p]^{(n-r)}dp$$

Example 4.5: One important special case of the PDF of the r-th order statistic is when the distribution for z is uniform. Then $F(z) = z$ and $f(z) = 1$ for $0 \le z \le 1$ and zero otherwise. In this case the PDF of the r-th order statistic, $z_{(r)}$, is

$$f_{(r)}(z_{(r)}) = \{n!/[(r-1)!(n-r)!]\}z_{(r)}^{(r-1)}(1-z_{(r)})^{(n-r)}, 0 \le z_{(r)} \le 1 .$$

This is a well-known distribution in statistics called a beta distribution. The quantile function for this distribution is numerically obtained and is denoted by BETAINV(p, r, $n + 1 - r$). It will be seen that the order statistics distribution rule above is simply an application of the uniform transformation rule applied to the uniform order statistics. The distribution has mean and variance given by

$$E(Z_{(r)}) = r/(n + 1), \quad V(Z_{(r)}) = [r(n - r + 1)]/[(n + 1)^2 (n + 2)].$$

Notice in passing that the values $x_{(r)}' = Q(r/(n + 1))$ divide all continuous distributions of X into $n + 1$ sections of equal probability $1/(n + 1)$

  Having obtained the distributions of the order statistics, the next step is to find the main statistical properties of these distributions. The mean of the r-th order statistic, $E(X_{(r)}) = \mu_{(r)}$, is referred to as the r-th **rankit**. Thus for the uniform distribution, $\mu_{(r)} = r/(n + 1)$. It can be shown that for the exponential distribution the rankits can be generated by a set of recurrence relations as follows:

$$\mu_{(0)} = 0,$$
$$\mu_{(1)} = \mu_{(0)} + 1/n$$
$$\mu_{(2)} = \mu_{(1)} + 1/(n - 1).$$

and in general

$$\mu_{(r)} = \mu_{(r-1)} + 1/(n + 1 - r).$$

  We will use the uniform and exponential rankits in later studies. Unfortunately, the rankits for most other distributions are not simple to calculate and for many distributions they exist as sets of tables scattered through the statistical literature. For a survey of such results, see Balakrishnan and Rao (1998). Using the uniform transformation rule and noting that the ordered $u_{(r)}$ lead to the corresponding ordered $x_{(r)}$ gives $x_{(r)} = Q(u_{(r)})$. We will show in Section 4.5 that, as a first approximation only, $E(\text{function}(Z)) = \text{function}(E(Z))$. Hence an approximate rankit is given by $E(Q(U_{(r)})) = Q(E(U_{(r)})) = Q(r/(n + 1))$. Over the years various studies have suggested $Q((r - 0.5)/n)$ gives an approximation that is better over a range of distributional models, which is why we have kept to this form for our previous uses. However, both forms are approximations and the use of expectations to represent the tail order statistics, which have highly skewed distributions, is not very helpful in interpreting quantile plots. Although the expectations of order statistics are difficult to handle, the percentile properties can be directly derived from the distribution rule. In particular, the median of the distribution of an order statistic, which we will call the **median rankit**, M_r, is given by the median rankit rule.

The median rankit rule

$$M_r = Q(\text{BETAINV}(0.5, r, n - r + 1)).$$

The quantity

$$p_r^* = \text{BETAINV}(0.5, r, n - r + 1),$$

we will call the **median-p_r**.

The actual distributions of the order statistics can be explored using percentiles. Figure 4.2 shows the 1st, 50th and 99th percentiles of the order statistics for a logistic distribution. It is clear from this that the distributions are skewed for order statistics from the tails of the logistic, skewed to the right in the right-hand tail and vice versa. In the central part of the distribution of X, the distributions of the order statistics are fairly symmetrical and the median values will be close to the means. The skewness of the distributions of the tail order statistics makes it difficult to judge quantile plots based on rankits. The mean of a skew distribution has no clear visual feature. However, with lines based on median rankits we look for the numbers of points above and below the line to be about half at all parts of the line.

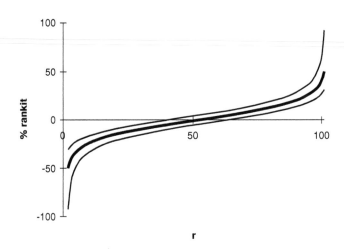

Figure 4.2. Percentile rankits for the order statistics of the logistic distribution

4.3   Transformation

In Section 3.2 we introduced the Q- and p-transformation rules. In this section we discuss these rules and their use in a little more detail. First, however, we need to prove the Q-transformation rule that the distribution of $Z = T(X)$ is $T(Q(p))$ when X has quantile function $Q(p)$. We restrict the function $T(X)$ to being a non-decreasing function of X. The distribution of Z can be expressed as the quantile function and CDF in the notation

$$z_p = Q_z(p) \text{ and } p = F_z(z_p).$$

The CDF can be written

$$p = \text{Prob}(Z \leq z_p)$$

which can then be written as

$$p = \text{Prob}(T(X) \leq T(x'))$$

where $z_p = T(x')$. But since $T(\ )$ is a non-decreasing function the inequality can be re-expressed as

$$p = \text{Prob}(X \leq x').$$

But this is exactly the expression that defines the p quantile, $x_p = Q(p)$, for the distribution of X, so $x' = x_p$. The transformation can therefore be expressed as $z_p = T(x_p)$ and therefore,

$$Q_z(p) = T(Q(p)).$$

It is seen then that the quantile function of the transformed variable, z, is found simply by transforming the quantile function of the original variable, x.

Example 4.6: Let X have an exponential distribution and Z be defined by $Z = X^2$. Thus

$$Q(p) = -\eta \ln(1 - p)$$

and therefore

$$Q_z(p) = [-\eta \ln(1 - p)]^2.$$

This distribution is a special case of the Weibull distribution.

In the above we have used an increasing transformation. Suppose $T(X)$ is a decreasing function, for example, $-X$ or $1/X$. The previous argument for increasing functions now has to be modified. We still have that

$$p = \text{Prob}(T(X) \leq T(x'))$$

where $Q_z(p) = T(x')$. But since $T(\ )$ is a decreasing function, the inequality can be re-expressed as

$$p = \text{Prob}(X > x') = 1 - \text{Prob}(X \leq x').$$

Hence

$$\text{Prob}(X \leq x') = 1 - p.$$

But this is exactly the expression that defines the $1 - p$ quantile, $Q(1 - p)$, for the distribution of X, so $x' = Q(1 - p)$. The transformation can therefore be expressed as $Q_z(p) = T(Q(1 - p))$. This thus gives the extended version of the Q-transformation rule for decreasing transformations.

Example 4.7: Suppose X has an exponential distribution and $Z = 1/X^2$. Then

$$Q_z(p) = 1/(-\ln p)^2.$$

This is a special case of a distribution called the Type 2 Extreme Value Distribution.

We showed in the p-transformation rule of Section 3.2 that if $Z = H(p)$ is a non-decreasing function with $H(0) = 0$ and $H(1) = 1$, then $Q_z(t) = Q(H(t))$ is a quantile function.

Example 4.8: Let X have an exponential distribution and the p-transformation be $p = t^2$, then

$$z_t = -\eta \ln(1 - t^2).$$

This is a special case of a distribution called the Burr Type X.

We now have two types of transformation. The Q-transformation allows us to find the distributions of new random variables obtained from, usually, increasing or decreasing functions of x. The p-transformation is a device for creating new and valid quantile distributions. Although they seem rather different operations mathematically, their function is to transform the two axes of the quantile plot.

Five specific cases of the use of transformations are of particular value in modelling:

(a) The uniform transformation rule that expresses any variable as $Q(U)$, where U is a uniform random variable, is a widely used property. For example, we have investigated the properties of the order statistics from basic definitions. We can also develop the properties using the uniform transformation rule. We imagine the order statistics generated from the order statistics of a uniform distribution, so $X_{(r)} = Q(U_{(r)})$. The properties of $X_{(r)}$ can then be studied using those of $U_{(r)}$.

(b) A special case of (a) is the transformation to the exponential distribution. This has a quantile function $= -\ln(1 - p)$. Hence the link to the uniform is given by

$$y = -\ln(1 - u), \text{ and in reverse } u = 1 - \exp(-y).$$

(c) If it is possible to transform from two distributions to the uniform, it is clear that it is possible to transform, via the uniform, from one to the other. In particular, it is possible to transform from a known distribution of x to the exponential distribution. Thus if $x = Q(u)$ and $y = -\ln(1 - u)$, then the transformation is

$$x = Q(1 - \exp(-y)) \text{ or } y = -\ln(1 - F(x)).$$

This result was originally shown by Renyi (1953).

(d) A set of transformations that have proved to be of considerable value in statistics are the Box–Cox transformations, (Box and Cox, (1964)), defined by

$$y = BC(z) = (z^\alpha - 1)/\alpha, \qquad \alpha \neq 0.$$
$$= \ln(z), \qquad \alpha = 0.$$

As α changes from negative, through zero, to positive values, the curves representing the transformation form a smoothly changing pattern. The log transformation is the limiting case as α approaches zero. Notice that the transformations refer to the general shape. Matters of position and scale can be dealt with as simple additional stages of transformation.

(e) In a wide range of situations the logs of the data have been used as the basis for modelling. Although this leads to simpler models it can cause problems when trying to work back to the underlying model for the raw data. In quantile function terms if $\lambda + \eta S(p)$ is the model for the log data, then the model for the raw data is simply $Q(p) = \exp[\lambda + \eta S(p)]$. In many situations λ will be a function of other, regressor variables. We will show later that there is no reason why this last model cannot be used directly to fit and analyze the data, without having to resort to using the log transformation on the data.

If we require the expectation of a Q-transformed distribution we have to integrate $T(Q(p))$ which is not in general the same as $T(\int_0^1 Q(p)dp)$, so the expectation of a function is not the function of the expectation, i.e., $E[T(X)] \neq T[E(X)]$. However, if we substitute $p = 0.25, 0.5, 0.75$, or any other specific value, p_0, we do have $Q_z(p_0) = T(Q(p_0))$, which gives a general quantile transformation rule, provided of course that $T(.)$ is a non-decreasing function. For example, we have the median transformation rule.

The median transformation rule

The median of a non-decreasing function is the function of the median,

$$Median[T(X)] = T[Median(X)].$$

We have already seen this general result for the specific case of the median rankit rule.

4.4   Simulation

Probably all spreadsheet software, all statistical software, all general-purpose languages and many pocket calculators provide the user with a simple way of generating **random numbers**. Although there are variations on the theme, the basic random number is a number in $(0,1)$ that represents an observation on a continuous uniform distribution. In quantile language the quantile function is

$$S(p) = p, \ 0 \leq p \leq 1,$$

and the p-PDF is

$$f_p(p) = 1, \ 0 \leq p \leq 1.$$

The random numbers generated by the software are not in fact truly random, they are pseudo-random, being generated by deterministic numerical algorithms. Often these algorithms require a starting value, a seed, from which they grow. In normal use a different seed is used each time a sequence of values is required. However, it is possible and sometimes useful to repeatedly generate the same stream of random numbers. Although the terminology varies, a common notation that we will use is to let a value given by such an algorithm be denoted by u, so we could write $u = RAND$. A stream of such uniform observations is denoted by $u_1, u_2, ..., u_n$. The generating mechanism is designed to produce a stream of approximately independent values. We refer to the generation of random variables in such a fashion, and also to the use of such values in the investigation of a model of any type, as **simulation**.

The uniform transformation rule of Section 3.2 showed that the values of x, from a distribution with quantile function $Q(p)$, can be simulated from

$$x_i = Q(u_i), \ i = 1, \ 2, \ ..., \ n.$$

The quantile function thus provides the natural way to simulate values for those distributions for which it is an explicit function of p.

Where a distribution is discrete or does not possess an explicit quantile function a wide variety of alternative approaches have been adopted. The only one that is of specific relevance to our subject matter is the simulation of the normal distribution. The most elegant way of obtaining simulated observations from a normal distribution is given by the Box–Muller formula, Box and Muller (1958). If u_1 and u_2 are independent random numbers and we use the transformations

$$x_1 = [-2 \ \ln u_1]^{1/2} \cos(2\pi u_2),$$

$$x_2 = [-2 \ \ln u_1]^{1/2} \sin(2\pi u_2),$$

then x_1 and x_2 are independent standard normal variates. We build on this idea in Chapter 13.

In a number of the applications of quantile distributions interest focuses particularly on the extreme observations in the tails of the data. To simulate these requires the simulation of the u_i, followed by the ordering of the u_i and thence the substitution into $x = Q(u)$ to obtain the ordered $x_{(i)}$. The non-decreasing nature of $Q(\ )$ ensures the proper ordering of the x. The process of ordering a large data set is very time consuming in relative terms even on a computer. Fortunately it is possible to simulate the observations in one tail without simulating the central values. We will state here how to do this. Consider the right-hand tail. The distribution of the largest observation has been shown to be $Q(p^{1/n})$. Thus the largest observation can be simulated by $x_{(n)} = Q(u_{(n)})$, where $u_{(n)} = v_n^{1/n}$ and v_n is a random number. If we now generate a set of transformed variables by

$$u_{(n)} = v_n^{1/n}$$

$$u_{(n-1)} = (v_{n-1})^{1/(n-1)} \cdot u_{(n)}$$

$$u_{(n-2)} = (v_{n-2})^{1/(n-2)} \cdot u_{(n-1)}$$

$$\text{etc.}$$

where the v_i, $i = n, n - 1, n - 2, \ldots$, are simply a simulated set of independent random uniform variables, not ordered in any way. It will be seen from their definitions that the $u_{(i)}$ form a decreasing series of values with $u_{(i-1)} < u_{(i)}$. In fact, the $u_{(i)}$ form an ordered sequence from a uniform distribution. Notice that once $u_{(n)}$ is obtained, the relations have the general form

$$u_{(m)} = (v_m)^{1/m} \cdot u_{(m+1)}, \quad m = n - 1, n - 2, \ldots.$$

The order statistics for the largest observations on X are then simulated by

$$x_{(n)} = Q(u_{(n)})$$
$$x_{(n-1)} = Q(u_{(n-1)})$$
$$x_{(n-2)} = Q(u_{(n-2)})$$
etc.

This method was introduced by Schucany (1972) and a similar result giving the order statistics in ascending order was given by Laurie and Hartley (1972). Notice that the nature of the uniform distribution means that each can be obtained from the other. Table 4.1 illustrates the simulation of the 4 largest values from a sample of 20 from a standard exponential distribution.

In general the statistical properties of functions calculated from the sample quantiles, such as the p-skewness index, are difficult to analyze theoretically. Simulation thus provides a practical procedure for studying behaviour. The simulated data is provided by $x_{(r)} = Q(u_{(r)})$. The quantile function used may take several forms:

(a) It may be a theoretical model, $Q(p)$. The parameters would be fixed within practical useful regions.

m	v	$v^{1/m}$	$u_{(m)}$	$-\ln(1 - u_{(m)})$
20	0.129	0.9027	0.9027	2.330
19	0.465	0.9605	0.8670	2.017
18	0.316	0.9380	0.8133	1.678
17	0.619	0.9722	0.7907	1.564

Note: Simulation of the four largest observations for a sample of 20 from an exponential.

Table 4.1. Simulation

(b) It may be that interest focuses on the statistics from the data-based fitted model. In this case $\hat{Q}(p)$ would be used in place of $Q(p)$.

(c) If there is doubt about the validity of the model, then a model-free approach can be used by using the empirical quantile function given by $\tilde{Q}_o(p)$, defined in Section 2.2. This will give a **bootstrap** sample, equivalent to sampling, with replacement, from the original data.

(d) If we wish to use an essentially model-free approach, but with a more continuous form of $Q(p)$, then the sample quantile function $\tilde{Q}(p)$, defined in Section 2.2, may be used. Unfortunately this only defines values for p inside $(1/n, (n-1)/n)$. For values in the small tail sections, a fitted distributional model is needed (see Hutson (2000)). If $\eta S(p)$ is the model, then the tail $Q(p)$ are given by:

$$\tilde{Q}_o(p) = \eta S(p) - \eta S(1/n) + x_{(1)},$$

or

$$\tilde{Q}_o(p) = \eta S(p) - \eta S((n-1)/n) + x_{(n)}.$$

The parameter η can be estimated by the method of percentiles using, for exampl'e, a wide $IPR(p)$.

In most simulation studies m samples of n observations are generated and the sample analyzes repeated m times to give an overall view of their behaviour. A technique that is sometimes used as an alternative to such simulation is to use a single sample of ideal **observations**, sometimes called a **profile** (see, for example, Mudholkar, Kollia, Lin and Patel (1991)). Such a set of ideal observations could be provided by the rankits, $E(X_{(r)})$, the approximation $Q((r-0.5)/n)$, or the median rankits, M_r, for $r = 1, ..., n$.

4.5 Approximation

Suppose $h(x)$ is some function of x and $h'(x)$, $h''(x)$, etc. are the first, second, etc. derivatives with respect to x. The value of x close to some specified value of x, say $x = a$, can be approximated by a Taylor series of powers of $(x - a)$

$$h(x) = h(a) + h'(a) (x - a) + h''(a) (x - a)^2/2! + \ldots$$

If $h(x)$ is a slowly changing smooth curve near a or if x is close to a so $(x - a)$ is small and decreasing with the power, then the first terms in this series will provide good approximations. We will be making use of three particular applications of this approximation and derive the basic results here.

(a) Approximate Expectations. Suppose we let x be a random variable, X, with $E(X) = a$ and take the expectation of $h(X)$. This gives

$$E[h(X)] = h(E(X))] + h'(E(X))E[(X - E(X))] \\ + h''(E(X)) E[(X - E(X))^2]/2! + \ldots$$

so

$$E[h(X)] = h[E(X)]$$

as a first approximation, using the fact that $E(\text{constant}) = \text{constant}$, and

$$E[h(x)] = h(E(X))] + h''(E(X)) V(X)/2!$$

as a second approximation.

(b) Approximate rankits. Suppose we replace x by $u_{(r)}$, the r-th order statistic from a uniform distribution which has a beta distribution (see Section 4.2), with parameters r and $n + 1 - r$, and also replace $h(\ )$ by a quantile function $Q(\ )$. We can use $a = r/(n + 1)$, which is the expectation of $U_{(r)}$. Hence,

$$Q(U_{(r)}) = Q(r/(n + 1)) + Q'(r/(n + 1))(U_{(r)} - r/(n + 1)) \\ + Q''(r/(n + 1)) (U_{(r)} - r/(n + 1))^2/2 + \ldots$$

From our look at order statistics and the nature of the quantile function it is clear that $Q(U_{(r)})$ is the distribution of the r-th order statistic of the distribution $Q(\ )$. Taking expectations of both sides gives the rankit for $X_{(r)}$ on the left-hand side. The second term on the right-hand side becomes zero, the third term has an expectation term which is the variance of $U_{(r)}$. This leads to

$$\text{rankit} = \mu_{(r)} = Q(r/(n + 1)) + Q''(r/(n + 1))$$
$$[(r(n - r + 1)]/2[(n + 1)^2(n + 2)] + \dots.$$

As a first approximation

$$\mu_{(r)} = Q(r/(n + 1)).$$

The approximation may obviously be improved by using the second and higher terms.

(c) Solving $Q(p) = x$ for p. Where distributions have an explicit CDF, $p = F(x)$, the value of p can be found from the observed x in a fitted model by using $F(x)$, with any parameters in it replaced by their estimated values. However, if there is no explicit $F(x)$ but an explicit and fitted quantile function, $\hat{Q}(p)$, a numerical solution has to be found to give the p for any x. Usually there is an ordered set of data, $x_{(r)}$, and the corresponding set of ordered $p_{(r)}$ is needed. These might be seen as the ordered set of uniform observations that, used with the fitted $\hat{Q}(p)$, would simulate the data exactly, i.e., $x_{(r)} = \hat{Q}(p_{(r)})$. Suppose p_0 is the current estimate of p for a given x. If we replace x by p in the Taylor series, put $h(p) = Q(p)$, the true quantile function for simplicity of notation, and use only the first two terms of the Taylor series, we have

$$Q(p) = Q(p_0) + Q'(p_0)(p - p_0).$$

Solving for p and using $x = Q(p)$ and $Q'(p) = q(p)$ gives as a better estimate:

$$p = p_0 + [x - Q(p_0)]/q(p_0).$$

If we are using the ordered data, the natural first estimate of $p_{(r)}$ is $r/(n + 1)$. The formula is used in an iterative fashion, with fitted $Q(p)$ and $q(p)$, until the given value of $Q(p)$ differs from x by less than some chosen small amount, depending on the accuracy required of the calculations. It should be noted that for p close to 0 or 1 the initial approximation may generate impossible values of p, less than 0 or greater than 1. This problem can be avoided by replacing such p by a or $1 - a$, respectively, where a is a very

small quantity, e.g., 0.00000001. The iteration then settles
to viable values. A problem that may be generated by this
is that in initial iterations the order of the $p_{(r)}$ may not be
correct. However, if $q(p)$ is a smooth function and n is
large, the transformation preserves an initially correct
ordering, and the iterations converge on the true values.
As most $Q(p)$ are smooth functions, four or five iterations
usually give accurate values for the p.

4.6   Correlation

In developing statistical models we are often interested in the relationships between variables. We consider the modelling of relationships in some detail in Chapters 12 and 13. It is useful, however, at
this stage just to introduce the idea behind a common measure of
relationship. Consider two variables, X and Y, that are unrelated; the
technical term is **independent**. It is intuitively reasonable that $E(XY)$
$= E(X)E(Y)$ for this situation. Suppose, however, that Y relates in a
linear fashion to X, so that large Y tend to occur with large X. This
will cause $E(XY)$ to increase above $E(X)E(Y)$. Conversely, if small Y
tend to occur with large X, then $E(XY)$ will reduce. The difference
produced by the relationship is thus defined by

$$C(X,\ Y) = E(XY) - E(X)E(Y).$$

This is called the **covariance**. It will be seen that if $X = Y$, then
the covariance becomes the common variance of X and Y, since $V(X)$
$= E(X^2) - E(X)^2$. It will be obvious from the definition that $C(X,\ Y)$
depends on the scale parameters of X and Y. To remove this dependence we standardize the covariance by dividing by the standard
deviations of X and Y. This leads to

$$\rho(X,\ Y) = C(X,\ Y)/[V(X)V(Y)]^{1/2}.$$

This is the **correlation coefficient**. It measures the strength of
the linear relation between X and Y. Listing its main properties:

(a)   If $Y = X$, then $\rho(X,\ Y) = 1$. If $Y = -X$, then $\rho(X,\ Y) = -1$.
(b) $-1 \le \rho(X,\ Y) \le 1$.
(c)   If X and Y are independent, the $\rho(X,\ Y) = 0$. The converse
      is, however, not true, since, for example, if there was a

perfect circular relation between X and Y we would still have $\rho(X, Y) = 0$.

(d) $\rho(X, Y)$ is independent of both position and scale. For example, linear transformations have no effect on the value of $\rho(X, Y)$. Thus

$$\rho(3 + 4X, 2 - 5Y) = \rho(X, Y).$$

(e) One special use of the correlation coefficient is between observed values X and their values, $Y = \hat{X}$, as predicted by a fitted model. Here we are looking for a large positive value to indicate a good fit, with a small value indicating a poor fit. The traditional measure of this is in fact the square of the correlation and is termed the **multiple correlation coefficient**, $R^2(X, \hat{X})$. Thus

$$R^2 = C^2(X, \hat{X}) / [V(X)V(\hat{X})].$$

(f) The sample values of correlation coefficients are derived by replacing the $E(\ )$, in the definitions of both $C(X, Y)$ and the $V(x)$ and $V(Y)$, by the corresponding sample averages.

As an illustration of the effects of correlation, Table 4.2 shows the population correlations between values of the ordered observations for a sample of 50 from a uniform distribution. It will be seen

				k			
r	1	2	3	4	5	10	20
1	0.700	0.566	0.485	0.429	0.387	0.270	0.169
5	0.903	0.827	0.764	0.712	0.668	0.511	0.336
10	0.942	0.890	0.844	0.803	0.765	0.615	0.413
15	0.955	0.913	0.874	0.838	0.804	0.658	0.436
20	0.960	0.922	0.886	0.852	0.819	0.672	0.421
25	0.962	0.925	0.889	0.854	0.820	0.663	0.358
30	0.960	0.921	0.883	0.845	0.808	0.627	0.169
35	0.955	0.910	0.865	0.820	0.776	0.540	
40	0.942	0.883	0.823	0.761	0.696	0.270	
49	0.700						

Correlation (U_r, U_{r+k}), $n = 50$

Table 4.2. Correlations between uniform order statistics

that these correlations are high and only die away slowly between observations that are further and further apart. The link between all distributions and the uniform distribution means that this high correlation is a universal feature of ordered data from any distribution. The visual consequence of this will be seen repeatedly in the illustrations in this book, for when we plot the ordered data the correlation causes it to show a snake-like shape. A particularly high value of one ordered observation will force up the values above it and vice versa. Such behaviour can easily cause us to look for further features in a model, when what we are seeing is just a natural consequence of this correlation.

In Section 1.3 a median-based measure of variability, MedAD, was introduced. This is defined as $\text{Median}(|X - M|)$. This is a more robust estimator than the variance, in that the sample statistic is unaffected by outlying observations in the tails of the data. A direct generalization of this for measuring covariance is the **comedian** defined by

$$\text{COM}(X, Y) = \text{Median}[(X - M_X)(Y - M_Y)].$$

This leads to a measure of correlation given by

$$\delta = \text{COM}/[\text{Med}AD_X \, \text{Med}AD_Y].$$

A detailed study of these measures is given in Falk (1997).

4.7   Tailweight

Tailweight and tail length are terms used to indicate the degree of probability in the tails of a distribution. This is reflected somewhat in the kurtosis, but is a measure of the peakedness or flatness of a distribution as well as the tail behaviour. Here we concentrate specifically on the behaviour of the extreme tails. In general, we refer to a heavy-tailed or long-tailed distribution as one having significant probabilities in the tail(s). If one uses, say, the normal distribution when the data actually comes from such a heavy-tailed distribution, then one will have a surplus of extreme observations. In applications such as quality control this will lead to inappropriate decisions. It is, therefore, useful to have some sample and population measures of tailweight. There have been a number of such measures suggested for application.

Using tail quantiles

One result of heavy tails is that for p in the outer tails $Q(p)$ will be particularly large or small. To assess this one needs to remove position and scale factors. Table 4.3 shows tail values of $Q(p)$ for a number of standardized distributions and parameter values.

A simple way of summarizing the tail length is to standardize, as suggested by Parzen (e.g., Parzen (1997)) and use the **identification quantile function:**

$$IQ(p) = [Q(p) - M]/2IQR$$

The values of $IQ(0.01)$ and $IQ(0.99)$ are compared with ± 1 and ± 0.5 to indicate tail lengths. For example, a distribution with $QI(0.99) > 1$ is regarded as having a long right tail, if it is between zero and 0.5 it is regarded as short tailed.

The TW(p) function

There have been a variety of proposals for ways of comparing and ranking distributions in terms of tailweight. One simple one, suitable for right-tailed or symmetric distributions, compares quantile functions as p approaches one. (See, for example, Hettmansperger and Keenan (1975).) As the value of $Q(p)$ will depend, at any p, on the effect of scale, the approach uses the relative rates of change of the quantile function rather than relative values. We use subscripts G and F to denote two distributions. The distribution G will have more probability in the tails than F, if the ratio $Q_G(p)/Q_F(p)$ is an increasing function of p for $p \geq 0.5$. This requires the derivative of the ratio to be greater than or equal to zero. Differentiating and sorting give

$$\text{slope} = (q_G Q_F - q_F Q_G)/Q_F^2 \geq 0.$$

This then leads to the rule:
   G will have heavier tails if

$$TW_G(p) \geq TW_F(p), \text{ where } TW(p) = q(p)/Q(p).$$

Example 4.9: For the exponential distribution

$$Q(p) = -\ln(1 - p), q(p) = 1/(1 - p), TW_E(p) = -1/[(1 - p)\ln(1 - p)].$$

Distribution	Parameter	M	UQ	Q(0.9)	Q(0.95)	Q(0.99)	Q(0.999)
Normal		0	0.5	0.95	1.22	1.72	2.29
Logistic		0	0.5	1.00	1.34	2.09	3.14
Symmetric Lambda	0.5	0	0.5	0.86	1.03	1.22	1.32
	−0.5	0	0.5	1.25	2.04	5.32	18.11
Cauchy		0	0.5	1.54	3.16	15.91	159.15
Extreme value 1		0	0.5	1.07	1.48	2.41	3.72
exponential		1	2.0	3.32	4.32	6.64	9.97
Weibull	0.5	1	1.41	1.82	2.08	2.58	3.16
	1.5	1	2.83	6.05	8.98	17.12	31.46
Power	0.5	1	1.22	1.34	1.38	1.41	1.41
	1.5	1	1.84	2.41	2.62	2.79	2.82
Pareto	0.5	1	1.41	2.24	3.16	7.07	22.36
	1	1	2.0	5.0	10.0	50.0	500
	1.5	1	3	11	32	354	11180

Table 4.3. Tail lengths of common distributions (standardized)

For the Pareto distribution

$$Q(p) = (1 - p)^{-\beta}, \; q(p) = \beta(1 - p)^{-\beta-1}, \; TW_P(p) = \beta/(1 - p).$$

Hence

$$TW_P(p)/TW_E(p) = -\ln(1 - p)/\beta.$$

This approaches infinity as p approaches one. Thus the Pareto has a heavier tail than the exponential.

Limiting distributions

It can be shown that as p approaches 1 the forms of many quantile functions can be approximated by $Q(p) \approx L(p)/(1 - p)^\beta$, where, for p approaching 1, $L(p)$ is slowly changing and approximately 1. For large β this is a heavy-tailed Pareto in the right tail. If the data has such a distribution, the quantile function for the log data will be

$$\ln Q(p) = \ln L(p) - \beta \ln(1 - p).$$

For p close to one this becomes

$$\ln Q(p) = -\beta \ln(1 - p),$$

which is an exponential distribution, for which the average of the observations gives an estimate of β. Notice that differentiating this relation gives:

$$q(p)/Q(p) = TW(p) = \beta/(1 - p),$$

using the result of the $TW(p)$ function. Thus the larger β the longer the tail, as we already know. Note also that for positive β, $TW(p)$ goes to infinity as p approaches one. A feature of the exponential distribution is that if we take some observation, $y_{(n-k)}$, as, say, a time measurement and use it as giving time zero; the distributions of the times from $y_{(n-k)}$ to later observations are also exactly the same exponential distribution. Thus the k-transformed observations $y_j = \ln x_{(n+1-j)} - \ln x_{(n-k)}$, $j = 1$ to k, are all observations on the exponential, and their average provides an estimate of β. This quantity is called the Hill estimator of β, the **tail index** (Hill (1975)).

Another use of the above results is that a plot of ln $x_{(r)}$ against $-\ln(1 - p_r)$ will tend towards a line of slope β to the right-hand end of the plot. See, as an example, Beirlant, Vynckier and Teugels (1996).

Short- or light-tailed distributions include the power and hence the uniform. Medium-tailed distributions are of the exponential type such as the Weibull and extreme distributions. Examples of long- or heavy-tailed distributions are the Pareto and Cauchy distributions.

4.8   Quantile models and generating models

It was shown in Section 4.4 that data from a distribution defined by its quantile function can be simulated by substituting a sequence of random numbers, denoted by $\{u_i\}$, from a uniform distribution into the quantile function, $Q(u)$. It follows from this that we can look at our models in two ways: first, as a mathematical relation between a p-quantile, X_p, and its quantile function, $Q(p)$, which is the quantile model; second, as a **generating model** which envisages the data sequence $\{x_i\}$ as generated by operating with the $Q(.)$ transformation on the sequence of independent uniform values $\{u_i\}$. The view of the distribution presented by the generating model is sometimes quite helpful in visualizing the situations described by quantile functions. The generating model is, however, not always equivalent to a quantile model, as is illustrated by the following example.

> **Example 4.10:** If U has a uniform distribution so has $1 - U$. Thus the two generating models $-\ln(1 - u)$ and $-\ln(u)$ both generate variables from the exponential distribution. However, only the first becomes a quantile model by putting $u = p$ as the second becomes a decreasing function of p.

The idea of a generating model can sometimes be helpful and its diagramatic form in a **block diagram** may also help in the visualization of distributions. We illustrate these ideas with two further examples:

> **Example 4.11:** A situation occurs in quality control studies where the output of a production process mixes products from two production lines having different statistical properties. Suppose the CDF of some dimension of the product is $F_i(x)$, where $i = 1$ or 2 depending on the production line. The probability of any individual item in the final output coming from line $I = 1$ is θ. Then using the rules of conditional probability, the CDF of the combined process, a **mixture distribution**, will be

$$F(x) = \theta \, F_1(x) + (1 - \theta) \, F_2(x).$$

Clearly there is no simple way of inverting this CDF to get the quantile function of the mixture distribution. However, we can represent the generating model of the process by representing the mixing process as a switch, as shown in the **block diagram** of Figure 4.3(a). If we consider the quantile model generated from the quantile functions for $F_1(x)$ and $F_2(x)$, using the addition rule, we have the linear model

$$Q(p) = \omega Q_1(p) + (1 - \omega)Q_2(p).$$

The generating model for this is shown in Figure 4.3(b). The difference between the two processes is clear. In the mixture situation each x is generated by one of two distinct mechanisms, the proportion of each being determined by the parameter θ. Mixture distributions should only be used where there is an evident mixing process in the data generation. In the linear model the whole process involves the influence of both the $Q_1(.)$ and $Q_2(.)$ mechanisms, the relative influence of each being determined by the parameter ω. The mixture model is sometimes proposed where there is no evident switching mechanism. It is evident that a quantile-based linear model is more likely to be appropriate in such situations.

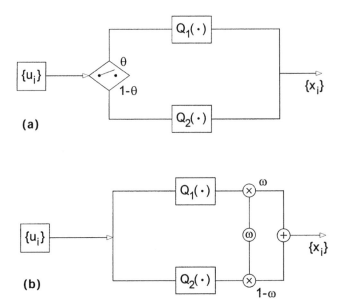

Figure 4.3. Examples of generating models. (a) Mixture distribution; (b) $\omega Q_1(p) + (1 - \omega)Q_2(p)$

Example 4.12: In Section 1.3 we mentioned the median absolute deviation:

$$\text{Med}AD = \text{Median}[\,|\,X - \text{Median}(X)|\,].$$

To obtain the sample value of this we subtract the sample median from each observation, drop the negative signs, reorder and take the median of this new data set. In population quantile terms we take our usual model with position and scale parameters and investigate the variable $y = |\,X - \text{Median}(X)|$, the absolute deviation. We can rewrite this as the generating model

$$\begin{aligned} y_i &= |Q(u_i) - Q(0.5)| \\ &= \eta[\,|S(u_i) - S(0.5)|\,]. \end{aligned}$$

We restrict ourselves to symmetric distributions, but even so this is not an increasing function of u over $0 \le u \le 1$ and so does not give the quantile model. Notice, however, that if we restrict u to $0.5 \le u \le 1$, then we do have an increasing function. Further, because of the symmetry of $|\,S(u_i) - S(0.5)|$ and the nature of the uniform distribution, we have the same relative spread of probability across the y values generated for the $0.5 \le u \le 1$ as for $0 \le u \le 1$. Thus the probability distribution given for y is the same whether u covers the whole range or just half the range. To get u to cover only half the range it may be generated from another uniform variable, V, using the transformation $U = (1 + V)/2$. The final generating model is thus

$$y = \eta[S\{(1+v_i)/2\} - S(0.5)],$$

which is now an increasing function of v over $0 \le v \le 1$ and is thus changeable to the quantile function:

$$Q_Y(p) = \eta[S\{(1 + p)/2\} - S(0.5)].$$

The median of this distribution is $\text{Med}AD = \eta[S(0.75) - S(0.5)]$.

4.9   Smoothing

On various occasions in our study we deal with quantities z_r, $r = 1, 2, 3, \ldots, n$, that vary randomly and whose variation can hide an underlying picture. The plots of the sample quantile and probability densities in Chapter 1 illustrate the problem. The simplest way to get at the

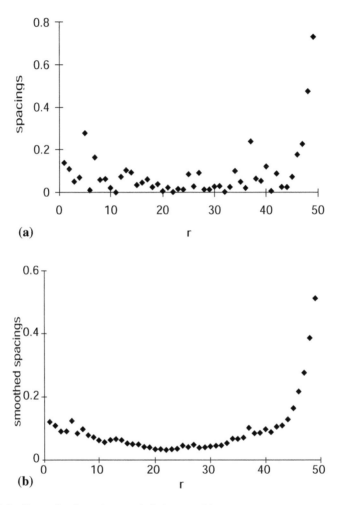

Figure 4.4. Example of spacings and their smoothing

underlying picture is by **smoothing**. There are a variety of approaches to smoothing, three of which we illustrate here. The first is based on the use of averages. Suppose we take a section of data containing, say five observations, and average the five. This average gives a smoothed estimate of the value of the series of z at the mid-point of the five. Thus if we consider the first five, then the smoothed value represents the smoothed series at z_3. If we drop z_1 and bring in z_6, the new average smoothes for z_4. We thus have a "moving" section of data and the consequent series of values are called the **moving average**, MA, of the series of z. If it is necessary to use an even number of observations in the section, the natural plotting point of the moving average is between two observations, e.g., for a four-term MA the point for the first four observations is at 2.5. For the next MA, it would be at 3.5. The solution is to centre the MA by averaging these two to get a value with plotting point 3. A problem with such an approach is that a choice has to be made of the width of the moving section to be used. Trial and error is a common solution.

Sometimes a single "badly behaved" observation has a considerable distorting affect on the previous methods. One way of reducing the distortion is to use the moving section but replace the mean by the median of the data in the section. This is called **median smoothing**.

An alternative approach is related to a classical method called **exponential smoothing**. Here we use a weighted average, $\tilde{z}_r = S_r/U_r$, where

$$S_r = z_r + az_{r-1} + a^2 z_{r-2} + \dots + a^{n-1} z_1,$$

and

$$U_r = 1 + a + a^2 + \dots + a^{n-1}.\ 0 \le a \le 1,$$

a being called the **smoothing constant**. This weighted average uses all the data up to z_r and puts the most weight on z_r and the observations close to it. Although it looks rather complex the numerator and denominator can both be calculated by using recurrence relations, thus

$$S_r = z_r + aS_{r-1},\ S_0 = 0,\ \text{and } U_r = 1 + aU_{r-1},\ U_0 = 0.$$

We will call $\tilde{z}_r$ the left smoothed value. If we start at z_n and work down we can obtain an exponentially smoothed value at z_r by parallel formulae. This is the right smoothed value. Averaging left and right smoothed values gives a combined smoothed value based on all the data, but weighting most the data close to z_r. Experience suggests that the value of a should be set quite high, above 0.9, to get a reasonable level of smoothing. The spreadsheet with its dynamic graphics capability allows for the value of a to be "tuned" to give a sufficiently smooth picture.

There is an important further approach called kernel smoothing. The methods are beyond the scope of this elementary text and are aimed primarily at creating good non-parametric pictures of the distributional shape. A survey paper by Ma and Robinson is in Balakrishnan and Rao (1998, Vol. 16). The simple methods discussed here are adequate to smooth the quantities of interest sufficiently for them to be compared with the equivalent population quantities.

4.10   Evaluating linear moments

L-moments and probability-weighted moments were introduced in Section 3.6. We concentrated there on definitions. Now we turn to issues of calculation. To evaluate the L-moments for a distribution we need expressions for the expectations in the definitions. We illustrate for λ_2 and quote the further formulae. The L-moment λ_2 was defined by

$$\lambda_2 = [E(X_{2:2}) - E(X_{1:2})]/2$$

From the results of Section 4.2 we have

$$E(X_{2:2}) = \int_0^1 Q_{2:2}(p)dp = \int_0^1 Q(p^{1/2})dp.$$

Substituting $z = p^{1/2}$ gives

$$E(X_{2:2}) = \int_0^1 Q(z) \cdot 2z\,dz.$$

A similar calculation with $E(X_{1:2})$ gives

$$E(X_{1:2}) = \int_0^1 Q_{1:2}(p)dp = \int_0^1 Q(1-(1-p)^{1/2})dp$$
$$= \int_0^1 Q(z')2(1-z')dz'.$$

The expectation of a sum or difference is just the sum or difference of the expectations, whether the variables are correlated or not, so on subtracting, simplifying and returning to p notation:

$$\lambda_2 = \int_0^1 Q(p)(2p-1)dp.$$

To find the formulae for the higher L-moments use is made of the general formula derived in Section 4.2, namely,

$$E(X_{r:n}) = \binom{n}{r}\int_0^1 Q(p)p^{r-1}(1-p)^{n-r}dp.$$

Substituting in the definitions for r and n and simplifying leads to

$$\lambda_3 = \int_0^1 Q(p)(6p^2 - 6p + 1)dp$$

and

$$\lambda_4 = \int_0^1 Q(p)(20p^3 - 30p^2 + 12p - 1)dp.$$

The right-hand side of all these expressions can be expanded as sums of probability-weighted moments, $\omega_{r,0}$. Thus, for example,

$$\lambda_2 = \int_0^1 Q(p)(2p - 1)dp = 2\int_0^1 Q(p)p\ dp - \int_0^1 Q(p)dp = 2\omega_{1,0} - \omega_{0,0}.$$

The first four L-moments in terms of $\omega_{r,0}$ are

$$\lambda_1 = \omega_{0,0}$$
$$\lambda_2 = 2\omega_{1,0} - \omega_{0,0}$$
$$\lambda_3 = 6\omega_{2,0} - 6\omega_{1,0} + \omega_{0,0}$$
$$\lambda_4 = 20\omega_{3,0} - 30\omega_{2,0} + 12\omega_{1,0} - \omega_{0,0}.$$

These relations between the population L-moments and the PWM can be used to obtain sample values for the L-moments in a simple direct fashion from the sample PWM, which, as seen in Section 3.6, can be obtained straightforwardly from the ordered data.

The linear form of the moments suggests that they might have particularly simple forms for linear models. This is in fact the case. For the general linear reflection form

$$Q(p) = \lambda + (\eta/2)[(1 + \delta)\ S(p) - (1 - \delta)S(1 - p)],$$

it can be shown that $\lambda_r = 2\eta\lambda_{S:r}$ for the even r and $\lambda_r = 2\eta\delta\lambda_{S:r}$ for odd r. Thus the L-coefficients of skewness and kurtosis become $\tau_3 = \delta\tau_{S:3}$

and $\tau_4 = \tau_{S:4}$, using an obvious notation. Thus it is seen once again that linear properties provide for simplicity and also that the δ parameter is a skewness parameter. Implicit also is that L-kurtosis arises from any shape parameters in $S(p)$.

4.11 Problems

1. Derive the probability-weighted moments, $\omega_{0,0}$ and $\omega_{0,1}$, for the distributions:

$$S(p) = 1/(1-p) - 1/p = (2p-1)/[p(1-p)].$$
$$S(p) = 1 - (1-p)^\beta.$$
$$S(p) = p/(1-p).$$

2. The rankit, $E(X_{(r)})$, is approximated by $Q(r/(n+1))$. Show that a rough identification for the first distribution of Question 1 is obtained by plotting $x_{(r)}$ against $(2r - n - 1)/[r(n + 1 - r)]$ for $r = 1, 2, ..., n$.

3. Suppose that X has the distribution

$$S(p) = (p^\alpha - 1)/\alpha - [(1 - p)^\beta - 1]/\beta$$
(a Tukey lambda distribution)

Obtain the distribution of the smallest observation from a sample of size n. Show that for large n this distribution is approximately exponential with threshold $1/\alpha$.

4. For the distribution of the previous question, relate the smallest observation from a sample of n to the corresponding smallest uniform observation. By expanding with Taylor's series show that, approximately,

$$x_{(1)} = [u_{(1)}^\alpha - 1]/\alpha + u_{(1)}.$$

5. The variable U has a uniform distribution with quantile function p. By examining the quantile function of $y = n(1 - u_{(n)})$ and using the limit that as $\beta \to 0$ the quantity $(z^\beta - 1)/\beta$ approaches $\ln z$, show that, for large n, y has approximately an exponential distribution.

6. Plot $p_r = (r - 0.5)/n$ against $p_r^* = \text{BETAINV}(0.5, r, n + 1 - r)$
    for $n = 10$ and $n = 100$, and consider the consequences for
    the interpretation of Q–Q plots using these alternatives.

7. Consider a variable X with quantile function $Q(p) = p/(1 - p)$. Show that
    (a) The distribution is of Pareto form with distributional
        range $(0, \infty)$.
    (b) The median of $Q(p)$, the lower quartile of the largest
        observation of a sample of two, and the upper quartile
        of the smallest observation of such a sample are all one.
    (c) The mean of the distribution is infinite. (In fact, all
        the ordinary moments are.)
    (d) Galton's skewness index is 1/2.
    (e) The reflected distribution has the form $(1 - p)/p$ and
        the log transformation gives the logistic distribution.
    (f) The distribution of $1/X$ is identical to the distribution
        of X.
    (g) The standard form of the skew distribution based on
        this model can be written as

    $$Q(p) = [(2p - 1) + \delta(2p^2 - 2p + 1)]/\{p(1 - p)\}.$$

    Hence show from first principles that the median is 2δ
    and the Galton skewness index is $\delta/2$.
    (h) For $p > 0.5$ the distribution does not have as heavy a
        tail as the distribution with quantile function $1/(1 - p)^2$.
    (i) The first three appropriate probability-weighted
        moments are

    $$\omega_{0,0} = \infty, \quad \omega_{0,1} = 1/2, \quad \omega_{0,2} = 1/6.$$

    (j) In general, the second L-moment relates to these
        PWM by

    $$\lambda_2 = w_{00} - 2w_{01},$$

    Hence show that although the PWMs, other than the
    one of zero order, are finite, the L-moments are not.

8. For the distribution in the previous question, simulate sam-
    ples of 10 and 50 observations and calculate the sample

values of the PWM of $Q7(i)$. Carry out a further simulation of the largest three values from 20 samples of 10.

9. The variable X has a standard exponential distribution. Show that as a first approximation the rankit for the largest observation of a sample of ten is 2.398. Show that a better approximation is 2.537 and that the median rankit from the actual distribution of the largest observation is 2.704.

10. Show that the general probability-weighted moment can be directly related to the expectations of order statistics by

$$\omega_{t, r, s} = \text{constant}^* \ E[X^t_{r+1, r+1+s}]$$

[See Landwehr and Matalas (1979) for some applications of this result.]

11. (a) By either numerical evaluation or series approximation for $T(p)$ show that $S(p) = p/(1 - p)$ is a long-tailed distribution but the logistic distribution, $S(p) = \ln[p/(1 - p)]$ is not long tailed.
    (b) Compare the logistic and the normal distribution by comparing values of $T(p)$ for p close to one.
    (c) For the distribution $S(p) = [1 - (-\ln p)^\beta]/\beta$ show that the distribution is of longer or shorter tail, relative to the type I extreme value distribution, depending on $\beta > 1$ or < 1, respectively.

12. An exceedence, y, is the time after some time u that an item survives, given that it is operational at time u, where $u = Q(p_0)$ and $Q(p)$ is the QF for the time to failure, X. Thus

$$F(y) = Pr(X - u \leq y \mid X > u).$$

Show that the quantile function of the variable Y is

$$Q[p_0 + (1 - p_0)p] - Q(p_0).$$

13. (a) Consider the sequence of polynomials, $P_r(p)$ defined by

$$P_1(p) = 1, \ P_2(p) = 2p - 1, \ P_3(p) = 6p^2 - 6p + 1,$$
$$P_4(p) = 20p^3 - 30p^2 + 12p - 1.$$

Show that $\int_0^1 P_r(p)dp = 0$ and that $P_r(1 - p) = -(-1)^r$ $P(p), \ r > 1.$

(b) Thus prove the result stated at the end of Section 4.10, that $\tau_3 = \delta\tau_{S:3}$ and $\tau_4 = \tau_{S:4}$ for the linear reflection model.

14. The Box–Cox transformation:

$$y = BC(z) = (z^\alpha - 1)/\alpha, \qquad \alpha \neq 0$$
$$= \ln(z), \qquad\qquad \alpha = 0$$

is often used to transform data to be a normal distribution. Show that if the distribution has quantile function $Q(p)$ then this implies that $Q(p) = [1 + \alpha\mu + \alpha\sigma N(p)]^{1/\alpha}$. Hence show that if from the data one derives $U_r = \ln[\tilde{f}(x_{(r)})/\phi_p(r/(n + 1))]$ and $V_r = \ln(x_{(r)})$, where $\phi_p(p)$ is the p-PDF for the standard normal and $\tilde{f}(x)$ is an empirical PDF for the data, e.g., based on Dp/Dx, then a plot of U_r against V_r corresponds to the line

$$U_r = -\ln(\sigma) + (\alpha - 1)V_r.$$

This data-based line can be used to judge the effectiveness of the transformation and also to estimate α. (See Parzen (1979) and Velilla (1993).)

CHAPTER 5

Foundation Distributions

5.1 Introduction

As with Meccano and Lego, our approach to modelling distributions is based on the principle of starting with a small number of standard parts. These are then added, or multiplied, together in different ways to create models of more complex structures. The complex structures are based on seeking to model sets of data. The building blocks that we will use are a number of simple statistical distributions. To use them to build more complex models we will develop various model building approaches. However, to carry out this model building we will need to express the building block models in quantile form. It should be noted that, as with Meccano and Lego, in this empirical approach to modelling our concern is that the final model behaves in ways that mimic the actual situation/data. This does not imply that the components of the model have any direct parallel within that situation/data. In this chapter we will look at the simplest component models in quantile form. For simplicity we will keep to models in basic form with $\lambda = 0$ and $\eta = 1$. In each section a summary table of properties is given. The moments, μ_r', are relatively simple for the distributions we will be studying and these moments are given in Tables 5.1 to 5.10. To calculate the moments about the mean from these the formulae in Table 3.2 are used.

5.2 The uniform distribution

As previously noted many pocket calculators and probably all statistical and spreadsheet software have an instruction RAND or RND. If this instruction is used repeatedly, it generates a series of numbers such as

0.669, 0.924, 0.740, 0.438, 0.631, 0.820, 0.144, 0.265, ...

These are randomly distributed numbers lying in the interval $(0,1)$. If the random variable, conventionally denoted by U, has such a distribution it is said to have a continuous uniform distribution in $(0,1)$, sometimes called a rectangular distribution. The probability density function is

$$f(u) = 1, \ 0 \le u \le 1$$
$$0 \text{ otherwise.}$$

The cumulative distribution function is

$$F(u) = u.$$

Reversing this to get the quantile function simply gives

$$Q(p) = p.$$

We thus have the simplest of quantile distributions. The RAND/RND instructions simply generate and simulate artificial, pseudo-random data from this distribution.

The uniform transformation rule of Chapter 3 shows that any distribution can be regarded as a transformed uniform distribution. The uniform is thus the foundation distribution of the building kit. The quantile and moment properties of the uniform are easily calculated and are presented in Table 5.1.

5.3   The reciprocal uniform distribution

Using the reciprocal rule it is evident that the quantile function for $1/U$ is

$$S(p) = 1/(1-p).$$

$S(p)$	p	Dist. Range $(0, 1)$
$F(u)$	u	
$s(p)$	1	
$f_p(p)$	1	
$f(u)$	1	$0 \le u \le 1$
$M = 0.5$	$IQR = 0.5$	$QD = G = 0$
$IPR(p) = 1 - 2p$	$T(p) = 2(1 - 2p)$	
$\mu_1 = 0.5$	$\mu_2 = 1/12$	$\mu_3 = 0$
$\mu_r' = 1/(1 + r)$		$\omega_{r0} = 1/(2 + r)$

Table 5.1. Distributional properties — uniform distribution

$S(p)$	$1/(1-p)$	Dist. Range $= (1, \infty)$
$F(z)$	$1 - 1/z$	
$s(p)$	$1/(1-p)^2$	
$f_p(p)$	$(1-p)^2$	
$f(z)$	$1/z^2$	$1 \le z \le \infty$
$M = 2$	$IQR = 2.666$	$QD = 1.333 \quad G = 0.5$
$IPR(p) = (1-2p)/\{p(1-p)\}$	$T(p){=}3(1-2p)/\{8p(1-p)\}$	
$\mu_r'\, r = 1, 2, \ldots$ do not exist	$\omega_{os} = 1/s$	

Table 5.2. Distributional properties — reciprocal uniform distribution

This distribution lies in the range $(1, \infty)$, seen by putting $p = 0$ and $p = 1$. The quantile density is $1/(1-p)^2$, so the p-PDF is $f_p(p){=}(1-p)^2$. The form of the distribution is that of a long-tailed decaying curve. This is not a distribution that often occurs as the basis for real data sets, but it is a basic building block of the modelling kit. Table 5.2 shows the main properties of the distribution.

5.4   The exponential distribution

The exponential distribution has been in use since the beginning of the 20th century to model the distribution of times to events. The events might be the arrival of telephone calls or the breakdown of equipment. If the events occur independently at random at the rate of η events per unit time, then the distribution of the time, x, from any defined moment to the next event has an exponential distribution. We used the exponential distribution several times in Chapter 1 to provide simple illustrations. Table 5.3 summarizes the main distributional properties of the unit exponential.

$S(p)$	$-\ln(1-p)$	Dist. Range $(0, \infty)$
$F(z)$	$1 - \exp(-z)$	
$s(p)$	$1/(1-p)$	
$f_p(p)$	$(1-p)$	
$f(z)$	$\exp(-z)$	$0 \le z \le \infty$
$M = \ln 2$	$IQR = \ln 3$	$QD = \ln(4/3)$
$G = 0.2619$	$T(p) = \ln[(1-p)/p]/\ln 3$	$G(p) = -[\ln\{p(1-p)\} + \ln 4]/\ln 3$
$\mu_1 = 1$	$\mu_2 = 1$	$\mu_3 = 2$
$\mu_4 = 9$		$\omega_{os} = 1/(s+1)^2$

Table 5.3. Distributional properties — the unit exponential distribution

5.5   The power distribution

The power distribution is rarely referred to in statistical texts as it rarely occurs in practice. However, it is one of the simplest forms of model and, therefore, can act as one of the building blocks for more complex models. In quantile form the standard power distribution is

$$S(p) = p^\beta. \ \beta > 0.$$

The main properties are obtained by simple calculations and are given in Table 5.4.

The parameter β is referred to as the power shape parameter. The moment properties are calculated from the QF in standard form as

The Expectation:

$$\mu_1 = \int_0^1 p^\beta dp = [p^{1+\beta}/(1+\beta)]_0^1$$

$$= 1/(1+\beta)$$

The Variance: best found by initially finding $E(X^2)$.

$$E(X^2) = \int_0^1 (p^\beta)^2 dp$$

$$= 1/(1+2\beta)$$

Hence

$$V(X) = 1/(1+2\beta) - [1/(1+\beta)]^2$$
$$V(X) = \beta^2/(1+2\beta)(1+\beta)^2$$

$S(p)$	p^β	Dist. Range $(0, 1)$
$F(z)$	$z^{1/\beta}$	
$s(p)$	$\beta p^{\beta-1}$	
$f_p(p)$	$(p^{1-\beta})/\beta$	
$f(z)$	$(z^{(1-\beta)/\beta})/\beta$	$0 \le z \le 1$
$M = 1/2^\beta$	$IQR = (3^\beta - 1)/4^\beta$	$QD = (1 - 2.2^\beta + 3^\beta)/4^\beta$
$\mu_1 = 1/(1+\beta)$	$\mu_r' = 1/(1 + r\,\beta)$	$T(p) = 4^\beta[(1-p)^\beta - p^\beta]/(3^\beta - 1)$
$\omega_{ro} = 1/(r + 1 + \beta)$		

Table 5.4. Distributional properties — power distribution, $\beta > 0$

A variant on the power distribution is its reflected but positive distribution. In its general form this is

$$Q(p) = \lambda + \eta[1 - (1 - p)^\beta].$$

This has a distributional range of $(\lambda, \lambda + \eta)$.

5.6   The Pareto distribution

The Pareto distribution was first used as a model for the distribution of income in a population. The quantile form is based on the standard QF of

$$S(p) = 1/(1 - p)^\beta, \ \beta > 0.$$

It is evident from the form of the model that from the reciprocal rule of Section 3.2 it is the distribution of the reciprocal of the power distribution. The reciprocal uniform is obviously a special case. The range of the distribution, from $p = 0$ and $p = 1$, is $(1, \infty)$. This can be changed to $(0, \infty)$ by setting $\lambda = -\eta$ in the general form. To find the PDF of the standard form, we differentiate, reciprocate and substitute to obtain

$$f(x) = (\gamma - 1)x^{-\gamma},$$

where

$$\gamma = (1 + \beta)/\beta.$$

Table 5.5 gives a summary of properties.

$S(p)$	$1/(1 - p)^\beta$	Dist. Range $(0, \infty)$
$s(p)$	$\beta/(1 - p)^{1 + \beta}$	
$f_p(p)$	$(1 - p)^{1 + \beta}/\beta$	
$f(z)$	$1/\{\beta z^{(1 + \beta)/\beta}\}$	$0 < z < \infty$
$M = 2^\beta$	$IQR = 4^\beta (1 - 1/3^\beta)$	$QD = 4^\beta (1 - 2/2^\beta + 1/3^\beta)$
$\mu_1 = 1/(1 - \beta), \ \beta < 1$	$\mu_r' = 1/(1 - r\beta), \ \beta < 1/r$	$T(p) = [1/p^\beta - 1/$
		$(1 - p)^\beta]/ \ 4^\beta(1 - 1/3^\beta)$
$\omega_{os} = 1/(s + 1 - \beta), \ s > \beta - 1$		

Table 5.5. Distributional properties — Pareto distribution, $\beta > 0$

5.7   The Weibull distribution

We have seen that the power and Pareto distributions are obtained as
the powers of the uniform and reciprocal uniform distribution. If we
take the positive power of the exponential distribution, we obtain the
Weibull distribution. Its quantile function in basic form is thus

$$S(p) = [-\ln(1 - p)]^\beta , \; \beta > 0.$$

This leads to a decaying PDF for $\beta \geq 1$ and a distribution with a
mode at $\eta(1 - \beta)^\beta$ for $\beta < 1$. This distribution was introduced first by
a physicist, Waloddi Weibull (1939), to model the distribution of breaking strengths of material. Since then the distribution has been very
widely used. The main properties of the distribution are presented in
Table 5.6.

5.8   The extreme type 1 distribution and the Cauchy distribution

We showed in Section 3.2 that the distribution of the largest of n
observations is $Q(p^{1/n})$. Thus each distribution has its own extreme
distribution. For many applications, such as the study of extremes of
weather, flood levels, and the like, the data records, and hence n,
increase steadily with time. It can be shown that the distributions of
the extremes tend to one of three limiting possible distributions. The
most common of these, the type 1 or Gumbel distribution, has the
standard quantile function:

$$S(p) = -\ln[-\ln(p)].$$

$S(p)$	$[-\ln(1 - p)]^\beta$	Dist. Range $(0, \infty)$
$F(z)$	$1 - \exp(-z^{1/\beta})$	
$s(p)$	$\beta[-\ln(1 - p)]^{\beta - 1}/(1 - p)$	
$f_p(p)$	$(1 - p)[-\ln(1 - p)]^{1 - \beta}/\beta$	
$f(z)$	$z^{1/\beta - 1} \exp(-z^{1/\beta})/\beta$	$0 \leq z \leq \infty$
$M = (\ln 2)^\beta$	$IQR = (\ln 4/3)^\beta - (\ln 4)^\beta$	$QD = (\ln 4/3)^\beta + (\ln 4)^\beta$
		$- 2(\ln 2)^\beta$
$\mu_1 = \Gamma(\beta + 1)$	$\mu_r' = \Gamma(\beta r + 1)$	$\omega_{r,0} = \Gamma(\beta + 1)/(r + 1)^{\beta + 1}$

Table 5.6. Distributional properties — the Weibull distribution

$S(p)$	$-\ln(-\ln p)$	Dist. Range $(-\infty, \infty)$
$F(z)$	$\exp[-\exp(-z)]$	
$s(p)$	$1/[p \ln p]$	
$f_p(p)$	$p \ln p$	
$f(z)$	$\exp(-z)[\exp(-\exp(-z))]$	$-\infty \leq z \leq \infty$
$M = -\ln(\ln 2)$	$IQR = \ln(\ln 4) - \ln(\ln 4/3)$	$QD = 0.1862$
$G = 0.1184$	$T(p) = 0.6359 \ln[\ln p/\ln(1-p)]$	
$\mu_1 = 0.57722$	$\mu_2 = 1.6449$	$\beta_1 = 1.2986$
$\beta_2 = 5.4$		

Table 5.7. Distributional properties — type 1 extreme value distribution

This has a distributional range of $(-\infty, \infty)$. The properties are summarized in Table 5.7 and the PDF shown in Figure 5.1.

The extreme distributions are by their very definition relatively long-tailed distributions. A very long-tailed distribution is the Cauchy distribution. This has been under investigation as a mathematical curve and a statistical distribution for some 300 years. It is a distribution with such heavy tails that none of the moments exist, i.e., all the integrals in their definitions become infinite. It has properties that cause it to be the counterexample and exception for many standard results and intuitions. For example, the average of a set of independent observations on a Cauchy distribution also has a Cauchy distribution

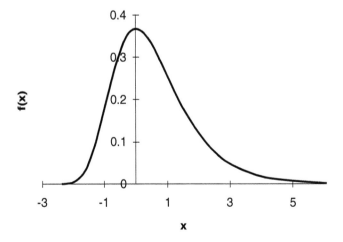

Figure 5.1. PDF of the extreme value (type 1) distribution

with the same parameters. What then is this strange model? The general quantile function is given by

$$Q(p) = \lambda + \eta \tan\left[\pi\,(p - 0.5)\right]$$

Thus the general form of the model includes position and scale parameters even although the moments do not exist. Remembering that a distribution can be simulated by replacing p by random uniforms, one interpretation of the distribution is that it is of the tangents of the angles of points placed at random round the right half of a unit circle. Table 5.8 shows the main properties of the distribution.

5.9   The sine distribution

It sometimes occurs that a plot of a set of data reveals several peaks, called modes, rather than one. This can often be explained as being due to a mixture of data from several different distributions. In this case a mixture distribution, based on summing CDF and PDF, should be used for modelling. However, if it is evident that the population is a genuine many-moded distribution, i.e., it is a **multimodal distribution**, then a many-peaked distribution is required. The sine distribution gives distributions with k peaks (see Figure 5.2). The basic quantile function is

$$S(p) = 2k\pi p + \beta\sin(2k\pi p); \ k = 1, 2, 3, \ldots; \ 0 < \beta < 1.$$

Although k is formally a parameter its integer value is simply set from a look at the data. The first term in the formula is required to ensure that the quantile function is a non-decreasing function. As it

$S(p)$	$\tan\left[\pi(p - 0.5)\right]$	Dist. Range $(-\infty, \infty)$
$F(z)$	$0.5 + (1/\pi)\tan^{-1}z$	
$s(p)$	$\pi[\sec^2\{\pi(p - 0.5)\}]$	
$f_p(p)$	$1/\pi\,[1 + \tan^2\{\pi(p - 0.5)\}]$	
$f(z)$	$1/[\pi\{1 + z^2\}]$	$-\infty \le z \le \infty$
$M = 0$	$IQR = 2$	$QD = 0$
$G = 0$	$T(p) = \tan\{\pi(0.5 - p)\}$	
μ_r do not exist		

Table 5.8. Distributional properties — Cauchy distribution

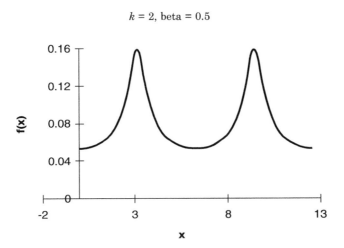

Figure 5.2. PDF of the sine distribution

is the uniform distribution its effect does not alter the basic symmetry of the PDF. The median of the distribution is $k\pi$. If we rewrite $\beta = (\alpha + 1)/(\alpha - 1)$, then α is the ratio maximum/minimum height of the distribution. The distributional range is $(0, 2k\pi)$. Table 5.9 gives a summary of properties of the distribution.

5.10 The normal and log-normal distributions

The normal distribution is the most commonly used distribution in statistics. Although it has no explicit quantile function, tables of its quantile function, usually called its probability integral and denoted by $\Phi^{-1}(p)$, have been in use for nearly 100 years (e.g., Sheppard (1903)). The 0.975 quantile of 1.96 is probably the best known number in statistics. Historically the distribution was first investigated in

$S(p)$	$2k\pi p + \beta\sin(2k\pi p)$
$s(p)$	$2k\pi[1 + \beta\cos(2k\pi p)]$
$DR(0, 2k\pi)$	max-f/min-f = $(1 + \beta)/(1 - \beta)$
	max points are at $p = (2i + 1)/2k$, $i = 0$,
	..., $k - 1$
$f_p(0) = f_p(1) = 1/2k\pi(1 + \beta)$	min at $p = i/k$, $i = 0, ..., k$

Table 5.9. Distributional properties — sine distribution

the 18th century. Tables of rankits were first published some 50 years ago (Teichroew (1956), Harter (1961)). We denote the standard normal quantile function by $N(p)$. This function is available via statistical and spreadsheet software. The general form for the normal has the form

$$Q(p) = \mu + \sigma N(p),$$

where μ is the mean and σ is the standard deviation. Paper for quantile plots based on the normal distribution, called **normal probability paper** or sometimes just probability paper, has been in regular use for many years, but the use of computer-generated plots is largely replacing the use of special graph papers. The main properties of the normal distribution are given in Table 5.10.

A major reason for the importance of the normal distribution is the fact that the distributions of many statistical quantities tend towards normality as the sample size increases. For example, it can be shown that the central order statistics tend to normality as the sample size increases (e.g., see Arnold, Balakrishnan and Nagaraja (1992)). To use this result we need the approximate means and variances of the order statistics. These come from results in Chapter 4. In Section 4.5 it was shown that if $X_{(r)}$ was the r-th order statistic, then the uniform transformation rule could be approximated by the linear relation:

$$X_{(r)} = Q(p_r) + Q'(p_r)\ (U_{(r)} - p_r)$$

where $p_r = r/(n + 1)$. The expectation of $X_{(r)}$ is, therefore, $Q(p_r)$, and the variance is given by $V(X_{(r)}) = q^2(p_r)V(U_{(r)})$, which using the formula for the variance of a order statistic of the uniform distribution from Section 4.2 gives for large n:

$$V(X_{(r)}) = q^2(p_r)p_r q_r/n.$$

$S(p)$	$N(p)$	
$f(z)$	$[1/\sqrt{(2\pi)}]\exp(-z^2/2)$	
Mean = 0	Variance = 1	Skewness = 0
$\mu_r = 0$	$\mu_r = (r - 1)(r - 3) \ldots 3.1$	$\beta_2 = 3$
r is odd	r even	
Median = 0	$IQR = 1.349$	

Table 5.10. Distributional properties — normal distribution

It can similarly be shown that the covariance of variables

$$Y_{(r)} = \sqrt{n}[X_{(r)} - Q(p_r)] \text{ and } Y_{(s)} = \sqrt{n}[X_{(s)} - Q(p_s)], \ r \leq s,$$

is given by

$$C(Y_{(r)}, Y_{(s)}) = q(p_r)q(p_s)p_rq_s.$$

By way of example if $p = 0.5$ and n is large and odd, $X_{(n/2 + 1)}$ is the sample median, m. We thus have that the sample median has the approximate distribution $N(M, q^2(0.5)/4n)$. Using the general values to obtain a standard normal gives that

$$Z_{(r)} = \sqrt{n} \cdot f_p(p_r)[X_{(r)} - Q(p_{(r)})]/\sqrt(p_rq_r)$$

has a standard normal distribution for large n. This approximation applies only for the central order statistics, for we know the tail order statistics have skew distributions.

It is occasionally useful to use the right-hand tail of a symmetric distribution as the basis for a right-tailed distribution, effectively folding the distribution in half and doubling the probability from the middle to any x_p. It will be seen that for the normal the quantile function of this **half normal** distribution is simply given ·by $N((p + 1)/2)$. Thus for $p = 0$ and $p = 0.5$ we get the values $N(0.5)$ and $N(0.75)$ as required. This distribution occurs in practice when we drop the negative signs from standardized normal data.

It sometimes occurs that the logarithm of the data has a normal distribution. In this case the data is said to have a **log-normal** distribution. In quantile terms this is expressed as

$$\ln(Q(p)) = \mu + \sigma \ N(p)$$

and hence

$$Q(p) = \exp[\mu + \sigma \ N(p)].$$

The distributional range is $(0, \infty)$. It can, however, occur that the minimum value is not zero but some value λ, giving rise to the more general quantile function:

$$Q(p) = \lambda + \exp[\mu + \sigma \ N(p)].$$

5.11   Problems

1.   Two forms of power distribution have been shown:

$$Q(p) = p^\beta \text{ and } Q(p) = 1 - (1 - p)^\beta.$$

By deriving and sketching the corresponding $f_p(p)$ show that these give four distinct distributional shapes depending on whether β is greater or less than one. For the second form above show that the probability-weighted moments can be obtained from

$$\omega_{os} = \beta/(s + 1)(s + 1 + \beta).$$

2.   The form of the Gumbel extreme value distribution used in this chapter has a positive mean. The reflected form is sometimes used. Show that the reflected distribution is just the log transformation of the exponential distribution.

3.   Show that if X is Cauchy, then so is $1/X$.

4.   Examine the symmetry of the sine distribution by showing that $Q(p) + Q(1 - p) = Q(1)$, for all p. Find a general expression for the IQR of this distribution as a function of k and β.

5.   Show that for a standard normal distribution the derivatives of the quantile function satisfy the relations

$$N'(p) = 1/\phi_p(p), N''(p) = N(p)/\phi_p(p), N'''(p) = N(p)/\phi_p(p)^2, \text{ etc.},$$

where $\phi_p(p)$ is the p-PDF for the standard normal.

6.   Examine the forms of the distributions of the largest and smallest observations from the following distributions:
     (a) The uniform distribution
     (b) The type 1 extreme value distribution $Q(p) = -\ln[-\ln p]$.
     (c) The type 2 extreme value distribution $Q(p) = [-\ln p]^{-\beta}$.
     (d) The type 3 extreme value distribution $Q(p) = -[-\ln p]^\beta$.

7. Use the approach described in Section 1.7 to explore the shapes of the PDF of the power, Pareto, Weibull and sine distributions for varying β.

CHAPTER 6

Distributional Model Building

6.1   Introduction

In the previous chapter we studied a range of common distributions. Many of these have been in existence for many years. Much of the development of distributional models has been based on seeking models with one or two parameters for fitting small samples of data. In these days of automated data collection one is often faced with very large data sets and the requirement for a small number of parameters can be relaxed. If there are a thousand observations it is probable that no two-parameter model will reasonably fit the data. We need to consider a larger catalogue of models and be able to build models that reflect the specific properties of the data being modelled. The objective of this chapter is to examine methods of building new models. In selecting approaches to be discussed, two considerations have been paramount. First, the models generated have structures that are likely to be useful to the practitioner. Thus we will be concerned with the forms of "model carpentry" that give commonly occurring tail shapes and meaningful parameters. Second, if the data comes from the given type of model it should be readily identified and validated. These two requirements put some realistic bounds on the types of model that ought to be considered and on the methods for constructing them. The models in the previous chapter provide a set of basic building blocks. In this chapter we therefore concentrate on how such simple components can be modified and combined to construct practical useful models.

6.2   Position and scale change — generalizing

The simplest form of model considered in most of this book can be written as $\lambda + \eta S(p)$. If $S(0)$ is not zero or $-\infty$, then it is sometimes convenient to shift the position by re-expressing the model as $\lambda - \eta S(0) + \eta S(p)$. This enables λ to be interpreted as the left-hand limit of the distributional range. Sometimes with a two-tailed distribution we would put $\lambda - \eta R(0.5) + \eta R(p)$, so that λ becomes the population median. This process is called shifting or centring.

> **Example 6.1:** The Pareto distribution, discussed in Section 5.6, has a distributional range of $(1, \infty)$. The centred Pareto can therefore be obtained as
>
> $$Q(p) = [1 / (1 - p)^\beta] - 1, \ \beta > 0.$$

An obvious and similar change can be obtained to adjust the scale of a distribution. Thus if we have a right-tailed distribution $\eta S(p)$ we may wish to have a fixed median of one. This is given by re-scaling the model as $S(p)/S(0.5)$.

Both the power distribution and the Pareto are defined for positive values of the shape parameter. It is useful to generalize these to cover all possible values of these parameters. To cover the values of $\beta = 0$ we make use of the mathematical limit that as $\beta \to 0$ the function $(p^\beta - 1)/\beta$ approaches the limit of $\ln(p)$. Forms that use this result to combine into one model a number of distributions we will call **canonical forms**.

> **Example 6.2:** Let us illustrate all the above three adjustments by considering the power distribution, p^β. This has a distribution over the range $(0, 1)$. If we change the position by first subtracting 1 and then the scale by division by β we make the distributional range $(-1/\beta, 0)$ and the quantile function is $(p^\beta - 1)/\beta$. Using the reflection rule to obtain a positive range we finally obtain
>
> $$S(p) = -((1 - p)^\beta - 1)/\beta,$$

which has distributional range $(0, 1/\beta)$. We thus have the power distribution shifted, scaled and reflected. If β now goes to zero this becomes the distribution $S(p) = -\ln(1 - p)$, the exponential distribution. To see what the distribution is for negative β let us write $\beta = -\alpha$, where α is positive. The distribution thus becomes $((1 - p)^{-\alpha} - 1)/\alpha$. It will be seen that this is a Pareto distribution with a shift of origin to give distributional range $(0, \infty)$ (Figure 6.1 illustrates). Approaching the model initially from a positive α gives a model often called the generalized Pareto distribution. It will be apparent that the generalized Pareto is equivalent to the generalized power. We keep to the generalized Pareto terminology as this is now a commonly used distribution.

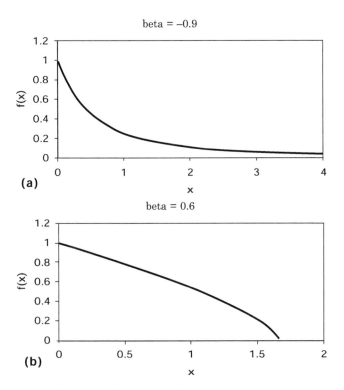

Figure 6.1. PDF of the generalized power distribution

Example 6.3: Consider the distribution

$$Q(p) = -((-\ln p)^\beta - 1)/\beta.$$

In the limit this is $-\ln[-\ln(p)]$, the type 1 extreme value distribution. For β positive it is the type 3 extreme value distribution and for negative parameter it is the type 2 extreme value distribution.

6.3   Using addition — linear and semi-linear models

In previous chapters we have illustrated a number of models con-structed by using the addition rule. These add basic parameter-free components, creating a linear function of the parameters. Such models are called **linear distributional models**. The most useful of these have the forms:

The one-parameter model $\qquad$ $Q(p) = \eta S(p)$

The two-parameter model $\qquad$ $Q(p) = \lambda + \eta S(p)$

The three-parameter model $\qquad$ $Q(p) = \lambda + \theta S_1(p)$
$$+ \phi S_2(p)$$

The three-parameter reflection model $\quad Q(p) = \lambda + \theta S(p)$
$$- \phi S(1 - p)$$

In its more convenient form this is re-expressed as

$$Q(p) = \lambda + (\eta/2)[(1 + \delta)\, S(p) - (1 - \delta)\, S(1 - p)]$$

Models with one non-linear parameter in $S(p)$, for example, a power of β, take the above forms with $S(p)$ replaced by $S(p;\beta)$. Models which are linear except for parameters in $S(p)$ we will term **semi-linear models**. The number of parameters becomes the total number of linear and non-linear parameters. For example, if in the last formula we have a single shape parameter in $S(p)$, but use different parameters for the original and reflected formulae, we will have a model with five parameters for position, scale, skewness, left-tail shape and right-tail shape, i.e., five meaningful parameters.

The properties of linear and semi-linear models are straightforwardly found, owing to the linearity associated with the operations of finding simple moments and percentile properties. If we have a general linear model

$$Q(p) = \lambda + \sum \eta_i\, S_i(p),$$

then the direct operations of calculating the expectation, $E[\ ]$, or finding the population median, $M[\ ]$, follow the form:

$$E[Q(p)] = \lambda + \sum \eta_i\, E[S_i(p)].$$

The operations that involve the subtraction of quantiles, such as $IQR[\ ]$, $IPR[\ ]$, $D[\ ]$, and $PD[\ ]$ follow the form:

$$IQR[Q(p)] = \sum \eta_i\, IQR[S_i(p)]$$

We also have

$$V[Q(p)] = \sum \sum \eta_i\, \eta_j\, C[S_i(p)\, S_j(p)],$$

where

$$C[S_i(p)\ S_j(p)] = E[S_i(p)\ S_j(p)] - E[S_i(p)]\ E[S_j(p)]$$

and

$$E[S_i(p)\ S_j(p)] = \int_0^1 S_i(p)\ S_j(p)dp$$

Appendix 1 gives some integrals of use here. For the sake of illustration consider the first two moments

$$Q(p) = \lambda + \eta_1\ S_1(p) + \eta_2\ S_2(p).$$

Using an obvious notation we have

$$\mu = \lambda + \eta_1\ \mu_1 + \eta_2\ \mu_2.$$

To obtain the variance we use the definition

$$\sigma^2 = \int_0^1 [(\lambda + \eta_1\ S_1(p) + \eta_2\ S_2(p)) - (\lambda + \eta_1\ \mu_1 + \eta_2\ \mu_2)]^2 dp$$

$$= \int_0^1 [\eta_1^2(S_1(p) - \mu_1)^2 + \eta_2^2(S_2(p) - \mu_2)^2$$

$$+ 2\eta_1\ \eta_2(S_1(p) - \mu_1)(S_2(p) - \mu_2)]dp$$

$$= \eta_1^2\sigma_1^2 + \eta_1^2\sigma_1^2 + 2\eta_1\ \eta_2\ \tau,$$

where

$$\tau = \int_0^1 (S_1(p) - \mu_1)(S_2(p) - \mu_2)dp$$

which simplifies to

$$\tau = \int_0^1 S_1(p)S_2(p)dp - \mu_1\mu_2 = \kappa - \mu_1\mu_2.$$

We will refer to τ and κ as the **standardized** and **direct quantile products**. These results refer to any forms of $S(p)$ irrespective of whether the $S(\ )$ involve further parameters.

All the L-moments and probability-weighted moments being inherently linear will clearly have the same simple additive property as the expectation.

It is sometimes of value to add two one-sided distributions with the same tail direction. We will consider two right-tailed distributions for the sake of illustration. Most of the basic component distributions in Table 1.2 are thus candidates. Consider a simple pairing,

$$Q(p) = \lambda + \eta_1 S_1(p) + \eta_2 S_2(p), \eta_1 > 0, \eta_2 > 0.$$

For most problems involving one-tailed distributions the population range is $(0,\infty)$ and so λ acts as the left-hand end of the distributional range of $Q(p)$, which for the present we will set at zero. Again it is convenient to alter the parameterization. We will use

$$Q(p) = \eta[(1 - \omega) S_1(p) + \omega S_2(p)], \eta > 0, 0 < \omega < 1.$$

Looking at the intermediate rule in Section 3.2 it is evident that $Q(p)$ lies between the two $S(p)$ distributions. The parameter ω controls the relative weight given to the two distributions. A useful generalization of this is obtained if ω is made an increasing function of p. Thus if ω is close to zero for small p and close to one for p close to one, then $Q(p)$ will behave like $S_1(p)$ to the left of the distribution, close to the origin, and like $S_2(p)$ in the right-hand tail. For example, if $\omega_1(p) = p$, then there is a steady shift from one distribution to the other as p increases. Sometimes it will be useful to obtain a fairly rapid shift from one model to the other. One weight function that achieves this is

$$\omega_2(p) = p^2(3 - 2p).$$

[an even faster shift is given by $\omega_3(p) = p^3(10 - 15p + 6p^2)$]. Note that the weights have a value of 0.5 at $p = 0.5$ and have $\omega(1 - p) = 1 - \omega(p)$. Figure 6.2 shows two distributions and their combinations using $\omega_2(p)$. It will be seen that the quantile function follows $Q_1(p)$ for about the lower third of the distribution and shifts towards $Q_2(p)$ for the final third.

In constructing the reflection family, the two $S(p)$ used naturally had comparable properties. This is not the case here. It therefore makes sense to force some comparability. The simplest method for parameter-free $S(p)$ is to use S that are standardized by

$$S^*(p) = S(p)/S(0.5).$$

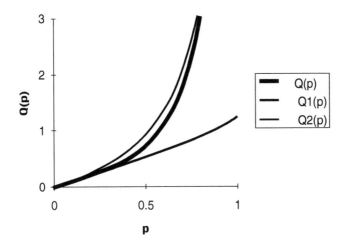

Figure 6.2. Adding two right-tailed distributions

Thus the distributions used have unit median. A consequence of this is that the constructed distribution, using any of the above weight functions, will also have unit median.

There are many ways in which the simple rules can be applied. We illustrate with a few examples.

Example 6.4: Two flexible right-tailed distributions are the two cases of the generalized Pareto introduced in Section 6.2 which has positive and negative parameters (but without scaling for unit median):

$$R(p) = [1 - (1 - p)^\alpha]/\alpha, \ \alpha > 0,$$

and

$$S(p) = -[1 - (1 - p)^{-\beta}]/\beta, \ \beta > 0$$
$$= [(1 - p)^{-\beta} - 1]/\beta.$$

Combining with constant weight and a scale parameter gives

$$Q(p) = \eta[\omega\{1 - (1 - p)^\alpha\}/\alpha + (1 - \omega)\{(1 - p)^{-\beta} - 1\}/\beta].$$

This distribution is used widely in the modelling of flood frequencies and is called the Wakeby distribution (Houghton (1978)). Figure 6.3 illustrates some of the forms of the distribution.

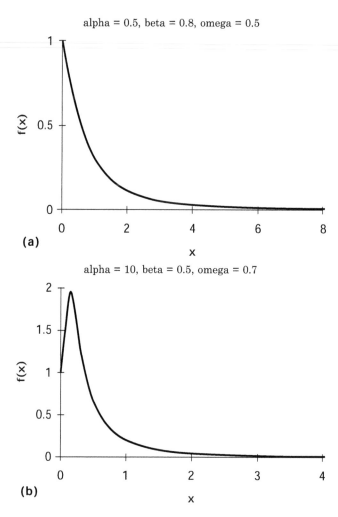

Figure 6.3. Forms of the Wakeby distribution

Example 6.5: A study of some earthquake data suggested that for low values on the Richter scale an exponential distribution described the distribution of the magnitudes of the shocks, whereas for more intense earthquakes the generalized Pareto gives a better description. A combination of the two using the second form of changing weights gave a good fit.

Example 6.6: In the discussion of the addition of quantile functions we have kept to models and combinations that inherently lead to valid increasing quantile functions. If we were to subtract distributions, the possibility arises of invalid models. Thus models involving such

constructions require careful design. An example of such a distribution is the Govindarajulu distribution, which involves subtracting
power distribution quantile functions:

$$Q(p) = (\beta + 1)p^\beta - \beta p^{\beta + 1}, \beta > 0.$$

The slope of $Q(p)$, given by $q(p)$, is

$$q(p) = \beta(\beta + 1)p^\beta[(1 - p)/p],$$

which is seen to consist of entirely non-negative terms. The distribution
is thus a valid distribution (see Govindarajulu (1977)).

Example 6.7: In the last chapter we briefly introduced the sine distribution to model multimodal distributions. One weakness of the model
is that $f_p(p)$ is not zero at the ends of the distribution. This can be readily
adjusted for by adding a symmetrical distribution, such as the logistic,
that has this property. This also has the effect of raising up the central
minima. However, for a bimodal distribution it gives a good shape as a
first model to study.

Thus the model is

$$Q(p) = 4\pi k p + \beta \sin(4\pi k p) + \gamma \ln(p/(1 - p)).$$

Figure 6.4 illustrates the distribution.

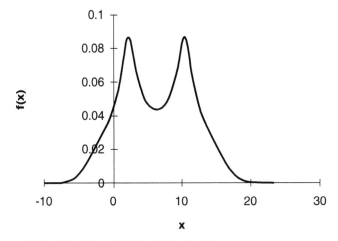

Figure 6.4. PDF of the sine and logistic distribution

6.4   Using multiplication

Using the simple component distributions the most natural pair to multiply together are the power and Pareto distributions. This leads to the four-parameter model with standard form:

$$Q(p) = p^{\alpha}/(1 - p)^{\beta}, \; \alpha, \; \beta > 0.$$

We will call this the **power–Pareto distribution**, $Po(\alpha) \times Pa(\beta)$. The distributional range is $(0, \infty)$. An example is given in Figure 6.5. To understand the shapes of the distribution, consider the quantile density function obtained by differentiation. This, after some simplification, is

$$q(p) = [p^{\alpha - 1}/(1 - p)^{\beta + 1}][\alpha + (\beta - \alpha)p],$$
$$= K(p)[\alpha(1 - p) + \beta].$$

The term in [ ] is positive for all p and the end values of $K(p)$ are given for all values of β by

$$
\begin{aligned}
K(0) &= \infty, & \alpha &< 1 \\
&= 1, & \alpha &= 1 \\
&= 0, & \alpha &> 1 \\
K(1) &= \infty, & \text{all } \alpha
\end{aligned}
$$

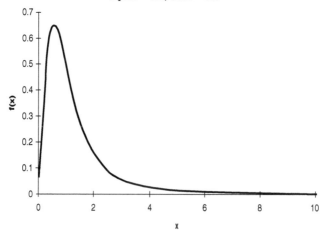

alpha = 0.5, beta = 0.5

Figure 6.5. PDF of the Power–Pareto distribution

As $f_p(p) = 1/q(p)$, it is clear that, irrespective of β, $f_p(p)$, and hence $f(x)$, is zero at $p = 1$ and also at $p = 0$ for the case $0 < \alpha < 1$. For $\alpha = 1$ $f(0)$ $= 1$ and for $\alpha > 1$, $f(0) = \infty$. Thus $\alpha = 1$ is the boundary between unimodal and decaying distributions.

The main percentiles are given by

$$LQ = 4^{\beta - \alpha}/3^\beta, \; M = 2^{\beta - \alpha}, \; UQ = 4^{\beta - \alpha} \cdot 3^\alpha,$$

6.5   Using Q-transformations

We now turn to the use of the Q-transformation rule of Section 2.2. An almost traditional procedure in statistical modelling has been the use of transformations of the data to obtain best fitting models. In quantile notation this leads to models of the form $T(x) = Q(p)$, where $T(\;)$ is a suitable transformation. In terms of the fit-observation plot we are transforming both x and p axes to seek a linear picture. An obvious problem with this is that one loses the natural dimension of the original data. Further, if the original data has a complex structure, for example, it may consist of a measured variable with added meas-urement error, $x + e$, or perhaps there may be other variables involved, then transforming the data just makes the difficulties worse. The natural approach with quantile functions is to develop a quantile model by transforming not the data but the quantile function. Thus we move from $Q(p)$ to $T^{-1}[Q(p)]$ and keep to the original data. The aim of this section is to illustrate how a simple set of transformations leads through a very wide range of distributions whose practical value over the years has been acknowledged by the giving of names. We focus on positive distributions with QF denoted by $R(p)$. It is convenient to introduce a simple notation for the main transformations that can be used as Q-transformations:

$$P(R(p)) = [R(p)]^\alpha \qquad \text{The Power Transformation,}$$

$$L(R(p)) = \ln(R(p)) \qquad \text{The Log Transformation,}$$

$$V(R(p)) = 1/R(1 - p) \qquad \text{The Reciprocal Transformation.}$$

With positive parameter α, these are increasing functions of p, for $0 \leq p \leq 1$. The generalized models that were discussed earlier in this chapter were designed to combine these in the form of

$$G(R(p)) = (R(p)^\alpha - 1)/\alpha \quad \text{The Generalizing Transformation.}$$

Two further transformations will be symbolised:

$$\text{Ref}(R(p)) \ \ = -R(1 - p) \quad \text{The Reflecting Transformation,}$$

$$C(R(p)) \ \ \ \ = \ R(p) - 1, \quad \text{A Centring Transformation.}$$

Having defined these functions we can systematically apply them to some simple basic distributions, expressed in standard form, and see what distributions emerge. Care needs to be taken to use the Q-transformation rule correctly for non-decreasing and non-increasing transformations. Some of the distributions obtained from transformations will be those already discussed. Others are distributions that have been used and named in the past. A few are not named distributions and we will refer to them by the symbols of the transformations that led to them.

The natural starting point is the uniform distribution $R(p) = p$. The first application of the transformations gives

Pp $S(p) = p^\alpha$ The Power Distribution,

Lp $S(p) = \ln p$ The Reflected Exponential Distribution,

Vp $S(p) = 1/(1 - p)$ The Reciprocal Uniform.

Having operated once on p, we may repeat the process, although with care and possible adjustment to ensure correctly increasing functions of p. We also sometimes simply return to a previous distribution, e.g., PPp is the same form as Pp. These cases we ignore. Thus we have

PVp $S(p) = 1/(1 - p)^\alpha$ The Pareto Distribution,

CVp $S(p) = p/(1 - p)$ A case of the Power–Pareto Distribution,

LVp $S(p) = -\ln(1 - p)$ The Exponential Distribution.

The centred Pareto based on the origin is

$CPVp$ $S(p) = (1 - (1 - p)^\alpha)/(1 - p)^\alpha$ The Centred Pareto.

The log transformation of CVp gives

$LCVp$ $\quad$ $S(p) = \ln[p/(1 - p)]$ $\quad$ The Logistic Distribution.

Raising the centred Pareto to a power gives the form:

$PCPVp$ $\quad$ $S(p) = [(1 - p)^{-\alpha} - 1]^{\beta}$ $\quad$ The Burr XII Distribution.

Applying repeat transformations to LVp, and for these model building approaches keeping α for the first parameter and β for the second parameter introduced, we have

$PLVp$ $\quad$ $S(p) = [-\ln(1 - p)]^{\alpha}$ $\quad$ Weibull Distribution,

$LLVp$ $\quad$ $S(p) = \ln[-\ln(1 - p)]$ $\quad$ The Reflected Type 1 Extreme Value Distribution,

$\text{Ref}LLVp$ $\quad$ $S(p) = -\ln(-\ln p)$ $\quad$ Type 1 Extreme Value Distribution,

$\text{Ref}PLVp$ $\quad$ $S(p) = -[-\ln p]^{\alpha}$ $\quad$ Type 3 Extreme Value Distribution,

$VPLVp$ $\quad$ $S(p) = [-\ln p]^{-\alpha}$ $\quad$ Type 2 Extreme Value Distribution.

$G\text{Ref}PLVp$ $\quad$ $S(p) = [1 - \{-\ln p\}^{\alpha}]/\alpha$ Generalized EV Distribution.

The essential feature to note here is that we have moved simply from one distribution to another through a wide range of named and useful distributions by a selection of very simple transformations.

Before leaving Q-transformations mention should be made of the exponential transformation $\exp(Q(p))$. This occurs if it is found that the logarithm of the variable x has the distribution $Q(p)$. The most used distribution of this type is the log-normal distribution, but there are also others, e.g., log-logistic.

6.6   Using p-transformations

The Q-transformation provides for transformation of the complete quantile function. The p-transformation, denoted here by a prefix p-, provides for transformation of the p term. For example, the transformation could be $Q(p^2)$. This provides a second means of transforming a quantile function. As $0 \leq p \leq 1$ a necessary requirement for the p-transformations is that they, too, are non-decreasing functions in the same range. This puts a severe limitation on the transformations available. Nonetheless, the available transformations open a wide range of distributional possibilities:

(a)   The Power p-Transformation

The simplest p-transformation is p^α which we will denote by Pp-. Note that if we apply Pp- to a distribution written in the previous notation it is not necessarily identical to using P as the very first Q-transformation applied. Thus Pp-$Q(p)$ is not the same as $Q(Pp)$. For example, Pp-LVp is $-\ln(1 - p^\alpha)$, but $LVPp$ is $-\alpha\ln(1 - p)$. Consider some examples:

Pp-LVp $\qquad\qquad\qquad\qquad$ $S(p) = -\ln(1 - p^\alpha)$.

Pp-PVp $\qquad\qquad\qquad\qquad$ $S(p) = 1/(1 - p^\alpha)^\beta$.

If this last distribution has β set at one we have the same result as ignoring the initial P transformation giving

Pp-Vp $\qquad\qquad\qquad\qquad$ $S(p) = 1/(1 - p^\alpha)$

Pp-CVp $\qquad\qquad\qquad\qquad$ $S(p) = p^\alpha/(1 - p^\alpha)$.

The Pp-CVp distribution relates to a family of distributions developed by Burr (1942). Thus

Pp-$LCVp = L(Pp$-$CVp)$ $\qquad$ $S(p) = \ln[p^\alpha/(1 - p^\alpha)]$.

This is the Burr II distribution, but is also sometimes referred to as the generalized logistic, for it is the p-transform of the logistic distribution.

Pp-$PCVp = P(Pp$-$CVp)$ $\qquad$ $S(p) = [p^\alpha/(1 - p^\alpha)]^\beta$

This is the Burr III distribution and in some literature is also called the kappa distribution.

A further member of the Burr Family is given by the reciprocal transformation:

$V(Pp\text{-}PCVp)$ $S(p) = [\{1 - (1 - p)^\alpha\}/(1 - p)^\alpha]^\beta$.
The Burr XII distribution.

Looking back at these models there are still one or two that can be further transformed. For example, the Weibull distribution, $PLVp$, has a p-transformation to $Pp\text{-}PLVp$. Thus

$Pp\text{-}PLVp$ $S(p) = [-\ln(1 - p^\alpha)]^\beta$.
The EW Distribution.

The EW is the Exponentiated Weibull Distribution which is used in reliability studies and will be mentioned again.

It is noted that one valuable p-transformation is where $\alpha = 1/n$, since as was seen in Section 3.2, $Q(p^{1/n})$ is the distribution of the largest observation of a sample of n.

(b)   The Reversed Power p-Transformation

A rather specialized p-transformation is given by using the reversed Power for $\alpha > 0$, thus

$$Rp\text{-}S(p) = S(1 - (1 - p)^\alpha).$$

This might be termed the Reversed Power p-transformation. One way of looking at this transformation is that the Pp-transformation has the effect of shifting probability to the right in $S(p)$ and Rp- shifts it to the left. For example, if α is 0.5, say, the median will be at $S(1/2^\alpha) = S(0.707)$ in $Pp\text{-}S(p)$, at $S(1/2)$ in $S(p)$, and at $S(1 - 1/2^\alpha) = S(0.293)$ in $Rp\text{-}S(p)$. There is a practical use of Rp- in finding the distribution of the smallest observation in a sample using $\alpha = 1/n$.

(c)   CDF p-Transformations

Suppose $F(v)$ is the cumulative distribution function of a continuous random variable, v, that lies in the distributional range (0,1), then $S(F(p))$ gives a viable p-transformation.

6.7   Distributions of largest and smallest observations

We saw in Section 3.2 that for a random sample of n from a distribution $Q(p)$ the distributions of the largest and smallest observations are

$$Q_{(n)}(p) = Q(p^{1/n}) \text{ and } Q_{(1)}(p) = Q(1 - (1 - p)^{1/n}).$$

As we have now introduced a number of distributions, it is of value to look briefly at the distributions of their extreme observations. One of the uses of knowing these distributions lies in the process of detecting outliers. An **outlier** is an observation that lies outside the natural range of sample values of a distribution due to the influence of some special cause, such as a copying or measurement error or perhaps due to contamination of the sample by data from a different population. If we take, say, the 99.5% quantile of $x_{(n)}$ and the 0.5% quantile of $x_{(1)}$, then we would be unlikely to see observations outside these limits. These limits are thus

$$Q(0.995^{1/n}) \text{ and } Q(1 - 0.995^{1/n}), \text{ respectively.}$$

This method provides only an approximation, since $x_{(1)}$ and $x_{(n)}$ are correlated; however, for large n the correlation is small. As the basis for a simple process of outlier detection the method is simple and effective. Often one is interested in only outliers at one end of the distribution. In this case the method is exact.

Let us now look at a number of cases where the distributions of the extreme observations have particularly simple forms. For the standard Weibull and exponential ($\beta = 1$) we have, after simplifying,

$$Q_{(n)}(p) = [-\ln(1 - p^{1/n})]^{\beta} \text{ and } Q_{(1)}(p) = (1/n^{\beta})[-\ln(1 - p)]^{\beta}.$$

Thus the largest observation has a distribution which is a power p-transformation of the Weibull and the smallest still has a Weibull distribution, but with a scale factor. For the largest observation on the power distribution and the smallest observation on the Pareto, we also get a return to the same distribution. Thus for the power distribution $Q_{(n)}(p) = p^{\beta/n}$ and for the Pareto:

$$Q_{(1)}(p) = 1/[1 - \{1 - (1 - p)^{1/n}\}]^{\beta} = 1/(1 - p)^{\beta/n}.$$

Thus the distributions are unaltered except that β becomes β/n.

Similar properties hold for the three extreme value distributions, where EV1 returns to an EV1 for the smallest observation, but with a change in the position parameter. The largest observations for the EV2 and EV3 return to their respective distributions with scale changes.

6.8   Conditionally modified models

Conditional probabilities

This short section is just a reminder that may be missed. Our use of expectations in defining correlation implied that we were not normally looking at deterministic relations between variables but rather at probabilistic links. In this section we will explore and illustrate some of the ways in which such probabilistic relationships may occur. We start from the basic probabilities of events. Suppose two experiments lead respectively to the outcomes A or not-A and B or not-B. If there were no relation between the A events and the B events, i.e., they are **independent**, intuition suggests that

Prob(A and B) = Prob(A) × Prob(B) (called the multiplication rule).

If A and B are related, then Prob(A) will in fact depend on the outcome B or not-B. Thus there is a need to replace Prob(A) and Prob(B) by the **conditional probabilities** such as Prob(A given that B has occurred), denoted by P(A| B). The formal definition of P(A| B) is given by

$$P(A| B) = P(A \text{ and } B)/P(B).$$

From this we can write

$$P(A \text{ and } B) = P(A| B) \times P(B) \text{ or } P(B| A) \times P(A)$$

This is the multiplication rule for non-independent events. Independence requires that $P(A | B) = P(A)$, which also implies that $P(B | A) = P(B)$. Thus the occurrence of one event does not influence the probabilities of the occurrence of the other. If we have non-independent events, then the unconditional probability of A is given by

$$P(A) = P(A| B)P(B) + P(A \qquad | \text{ not-B})P(\text{not-B}).$$

P(A) is thus the total of the probabilities that lead to A. If B has more than two possible outcomes, denoted by B_i, $I = 1, 2, ..., k$. Then this relation generalizes to

$$P(A) = \Sigma_{i=1}^{k} P(A|B_i)P(B_i)$$

which is sometimes called the **chain rule**.

The circumstances in which distributions arise sometimes involve features that create the need to use conditional probability ideas to modify the quantile function of the model.

Blipped distributions

There is no proper name for the models to be described here but the title gives a rough indication of the situation. Consider the following illustrative example.

> **Example 6.8:** The exponential distribution often models the time to failure, t, of an item of equipment. The CDF is $F(t) = 1 - e^{-\lambda t}$. For some items of equipment there is a distinct possibility, probability P, of failure when the equipment is first switched on. Only if it survives this does the exponential apply. Thus the probability, p, of failure up to time t is given by the chain rule as
>
> $$p = P + (1 - P)(1 - e^{-\lambda t}).$$
>
> The quantile function has to be zero for $p < P$, since with this probability the equipment fails at time zero. For $p \geq P$ we obtain $Q(p)$ by solving the last expression for t giving
>
> $$Q(p) = -(1/\lambda)\ln[1 - \{(p-P)/(1-P)\}], \qquad p > P.$$
> $$= 0, \qquad\qquad\qquad\qquad\qquad\qquad\qquad p \leq P.$$

The occurrence of a "blip" in the probability distribution at zero is a relatively common feature. The argument of the example holds generally so that if the basic quantile function is $Q(p)$ the modified model is

$$Q_m(p) = Q\{(p-P)/(1-P)\}, \qquad p > P.$$
$$= 0, \qquad\qquad\qquad\qquad\qquad p \leq P.$$

Truncated distributions

It sometimes happens that we are unable to observe data outside a fixed range of values of the variable. For example, a measuring device for earthquakes may record all shocks over 3 on the Richter scale. It will provide no information on either the number or the magnitude of shocks of less than three. This phenomenon is called **truncating**. It is clearly a case of conditional distributions. Our example is of truncating to the left. If we have no data above a certain value, c, then this is called truncating to the right. This will occur, for example, when a trial of the lifetimes of items under stress is stopped at some time before all items have failed. In this latter situation, denote the CDF of the truncated distribution by $F_T(x)$ and the non-truncated distribution by $F(x)$ $(= p_u)$. The conditional CDF is $F_T(x) = \text{Prob}(X \leq x \mid x \leq c)$. From the definition this is

$$F_T(x) = F(x \text{ and } x \leq c)/F(c)$$
$$= p_T = F(x)/F(c), \qquad x \leq c.$$

Denoting $F(c) = p_c$ this gives $p_u = p_T p_c$ for the corresponding non-truncated probability. Denoting the quantile function for the truncated distribution by $Q_T(p_T)$ we see that the p_T quantile, x, of the truncated distribution can be expressed as $Q_T(p_T)$ or as $Q(p_u) = Q(p_T p_c)$. Thus dropping the T subscript for p, we have the final form for the quantile function of the truncated distribution:

$$Q_T(p) = Q(p \, p_c).$$

It is seen that, as required, the distributional range is $(Q(0), Q(p_c))$. For a left truncated distribution the corresponding result is

$$Q_T(p) = Q[p_c + p(1 - p_c)].$$

This has distributional range $(Q(p_c), Q(1))$.

Example 6.9: Suppose x represents the speed of particles with an exponential distribution measured by a device that is unable to measure speeds greater than c (not the speed of light!), then $p_c = 1 - e^{-\lambda c}$ and hence

$$Q(p) = -(1/\lambda) \ln[1 - p(1 - e^{-\lambda c})], \, x \leq c.$$

Example 6.10: If a distribution that is symmetrical about zero is truncated at zero leaving only the positive values, its distribution will be given by

$$S_T(p) = S[(1 + p)/2].$$

Such distributions are called **half-distributions**. The most common examples are the half-normal and the half-logistic. The applications of these are situations where the interpretation of the data does not depend on sign, but only on magnitude. Thus signs are dropped and half-distributions used.

Censored data

In truncation, we have no information at all outside the truncated range. It sometimes happens that we do have some information. For example, suppose in a test of fatigue for metal sheets, a set of 20 sheets is subject to the same conditions of high stress. After 3 months the experiment is stopped with 14 plates having fractured. We thus have 14 observations. However, we also know that six plates would have fracture times greater than 3 months, even although we do not know the values and never will. This type of data is called **censored** data. There are two central variables in censoring: the value of x, such as time, and the number of observations for which we have values. A consequence of this is that there are two possible types of censoring. In type I censoring, the right or left limit of x, $x = c$, is fixed and the number of observations, R, explicitly obtained is a random variable. The fatigue situation as described involved type I censoring. In the example, it turned out that $R = 14$. It may happen that we fix $R = k$ as the basis for the censoring. For example, we could have decided to stop the experiment as soon as 14 plates had fractured, which might have happened after 7 weeks. This approach is called type II censoring. In this situation the length of the experiment is the random variable. The k observations correspond to the quantile function $Q(p_r)$ $r = 1, ..., k$, where p_r still depends on n. The expected duration of the experiment, $E(X)$, will be the expected value of $X_{(k)}$, given approximately by $Q(k/(n + 1))$, since it is the occurrence of $X_{(k)}$ that terminates the experiment.

6.9   Conceptual model building

It is sometimes possible to argue conceptually for a particular model, at least as a first trial model. This is best shown by illustration.

Example 6.11: The sales of a consumer durable such as a new type of TV or computer will depend on some general factor related to people's willingness to invest (measured by a constant, parameter, α) and also on the peer pressure of others who already have the item. If at time t a proportion $p(t)$ of the population has the item, then the peer pressure could be represented by $\beta p(t)$. Adding these together, the probability of a purchase in a small interval δt at time t by those not yet having made a purchase could be written as $[\alpha + \beta \, p(t)]\delta t$. This is the probability of buying in time δt, which has PDF $f(t)$, but with the condition that they have not yet bought $(1 - p(t))$. This conditional probability is thus $[f(t) \, \delta t]/(1 - p(t))$. Equating the two expressions gives

$$f(t) = f(Q(p)) = [\alpha + \beta \, p(t)][1 - p(t)]$$

Inverting to get the quantile density function, $q(p)$, gives

$$q(p) = 1/[\{\alpha + \beta \, p(t)\}\{1 - p(t)\}]$$
$$= [1/(\alpha + \beta)][1/\{1 - p\} + \beta/\{\alpha + \beta p\}]$$

Integrating to get the quantile function gives

$$Q(p) = [1/(\alpha + \beta)][-\ln(1 - p\} + \ln\{\alpha + \beta p\}].$$

Thus the distribution is the sum of an exponential and a more general distribution that lies in the range $(\ln(\alpha), \ln(\alpha + \beta))$. If we repeat the calculation with $\beta = 0$, we obtain the exponential distribution with $\eta = 1/\alpha$. Thus we have a conceptual justification for the exponential as the distribution that arises when there is a constant conditional probability of an event immediately occurring, given that it has not yet occurred.

Example 6.12: A set of data on summer humidity at a weather station in Sheffield, UK, is to be modelled. Humidity is measured as a percentage $(0, 100)$. The nature of the local climate is that it is rarely very dry, indeed rarely less than 50% humidity. Most of the time the humidity is in the 70s and 80s with occasional 90s. The situation thus implies a negative skewness with an upper threshold at 100. If we had a low threshold at zero and a single peaked distribution with a longish tail to the right, the Weibull would probably be a natural first model to try. By using the reflection rule, a reflected Weibull distribution with an upper threshold of 100 is given by

$$Q(p) = 100 - \eta[-\ln(p)]^{\beta}.$$

The distributional range of this is in theory $(-\infty, 100)$; however, with the choice of parameters needed to get a reasonable fit, the probability of negative fitted humidity is negligible.

6.10   Problems

1.   Express the Weibull distribution in canonical form. Show
     that for β becoming zero the distribution becomes a
     reflected extreme value type 1 distribution. For the case
     where $\beta = -\gamma$, $\gamma > 0$, show that (a) the distributional range
     is $(-\infty, 1/\gamma)$, and (b) $Q(p_0) = 0$ for $p_0 = 1 - 1/e$. Plot the form
     of this distribution for $\gamma = 0.5$. Consider the relation
     between this distribution and the generalized extreme
     value distribution considered in Section 6.2.

2.   If X has the Power–Pareto distribution, denoted by
     $P \times P(\alpha, \beta)$, show that the distribution of $1/X$ will be
     $P \times P(\beta, \alpha)$.

3.   Construct a reflection model with parameters (λ, η, δ) with
     the distribution $S(p) = 1/\sqrt{(1-p)}$. Calculate its main
     quantile properties.

4.   Investigate the construction and quantile properties of the
     following distributions:
     (a) $S(p) = (2p - 1)/[p(1 - p)] + 2\ln[p/(1 - p)]$.
     (b) $S(p) = p + (\beta/2k)(2p - 1)^k$.
     (c) $S(p) = -p\ln(1 - p)$.
     (d) $S(p) = p^2(3 - 2p)$.
     (e) $S(p) = -\ln(1 - p^\alpha)$, considering $\alpha = 0.5$ and $\alpha = 2$.

5.   Using the results of Section 6.3, show that for the three-
     parameter reflection model $\mu = \lambda + \eta\delta$ and $\sigma^2 =$
     $(\eta^2/2)[(1 + \delta^2)\sigma_S^2 + (1 - \delta^2)\tau]$, where τ is based on $S(p)$ and
     $-S(1 - p)$. Show specifically that for the skew logistic dis-
     tribution $\sigma^2 = (\eta/2)^2[\pi^2/3 + \delta^2(4 - \pi^2/3)]$. [Note that from the
     table of integrals $\kappa = \pi^2/6 - 2$.]

6.   Treating the various Q-transformations as operators, con-
     sider how many possible theoretical transformations there
     are of the forms PV, LPV, PLPV. Show that if $Q(p)$ is a
     positive distribution, there are a number of equivalencies
     among these, for example, $PP = P$; $VV = I$, the identity
     operator; $LP = L$ (in form ignoring scale); $VPV = P$ $PVP =$
     $VP = PV$; $VCV = CV$. A consequence of these is that many

of the different looking transformations lead to the same distributions.

7. The PDF of a right-tailed distribution has a value of α at $x = 0$ and tends to zero at infinity. One possible form would have p-PDF of

$$f_p(p) = (1 - p)(\alpha + \beta p).$$

Find the quantile function and consider its properties.

8. A measure of the severity of an illness has a distribution $Q(p) = \lambda + \eta S(p)$ over a population of those diagnosed as having the illness. A new treatment is to be tried. What modified models might be appropriate to the treated population if:
   (a) All respond in the same way by a constant reduction in severity?
   (b) The treatment has an effect only on the most severely ill half of the population?
   (c) The treatment works for all but in general has an effect which increases with the severity of the illness?

CHAPTER 7

Further Distributions

7.1   Introduction

In Chapter 5 a number of distributional models were introduced. These were selected for their simplicity and common use and also for their ability to act as components for the construction of more complex distributions. In this chapter we examine a number of distributional families that exemplify the building of complex models from simple ones. These will include some distributions that have explicit quantile functions but not explicit CDF or PDF. We also take the opportunity to comment on discrete distributions and other statistically important distributions that do not have explicit quantile functions.

7.2   The logistic distributions

In previous chapters the logistic and skew logistic distributions have been used repeatedly to illustrate a range of calculations. Here the main features are summarized for completeness. The book by Balakrishnan (1992) gives a comprehensive coverage of the symmetric distribution. The standard symmetric logistic distribution is defined in terms of its quantile function:

$$S(p) = \ln(p/(1 - p)).$$

Table 7.1 summarizes the distributions properties.

The logistic model has been widely used both as a model for data and as a simple model for illustrative purposes, e.g., Cox and Hinkley (1974).

If we use the parameterization of Section 4.6 we can weight the reflected exponential and exponential distributions unequally to give a skew logistic distribution. This is

$$Q(p) = \lambda + (\eta/2). \, [(1 + \delta)(-\ln(1 - p)) + (1 - \delta) \ln p]. \, -1 \leq \delta \leq 1.$$

$S(p)$	$\ln[p/(1-p)]$	
$F(z)$	$1/(1+e^{-z})$	
$s(p)$	$1/[p(1-p)]$	
$f_p(p)$	$p(1-p)$	
$f(z)$	$e^{-z}(1-e^{-z})^{-2}$	
$M = 0$	$IQR = \ln3$	$QD = G = 0$
	$T(p)=\ln[(1-p)/p]/\ln3$	
$\mu_1 = 0$	$\mu_2 = \pi^2/3$	$\mu_3 = 0$
$\mu_4 = 7\pi^4/15$	$\omega_{10} = 0.5$	$\omega_{20} = 0.5$

Table 7.1. Distributional properties — the standard logistic distribution

This uses weights that sum to one. Rewriting the model gives

$$Q(p) = \lambda + (\eta/2). \ [-\delta \ln p(1-p)) + \ln\{p/(1-p)\}],$$

which shows the effect of the skewness coefficient, δ, on the distribution. Note that this form leads to a scale parameter of $\eta/2$ for the symmetric form, with $\delta = 0$. The main properties of the skew logistic distribution are shown in Table 7.2 and are derivable by simple algebra and calculus.

7.3   The lambda distributions

The lambda distributions were originally developed as formulae for transforming uniform random numbers to simulate new distributions with a rich variety of shapes; for example, see Hastings et al. (1947), and Tukey (1962). Their use as models for data only came later. The family has been used for a range of different applications, for example, air pollution, Okur (1988); climate studies, Abouammoh and Ozturk (1987); finance, McNichols (1987); and inventory modelling, Nahmris (1994).

We will study the family of distribution under four forms which, including the position and scale parameters, have three, four or five

$S(p)$	$[(1 + \delta)(-\ln(1-p))$	
	$+ (1 - \delta) \ln p]/2$	
$s(p)$	$(1 - \delta + 2\delta p)/[2p(1-p)]$	
$f_p(p)$	$2p(1-p)/(1-\delta+2\delta p)$	
$M = \delta \ln2$	$IQR = \ln3$	$QD = \delta \ln(4/3)$
$G = 0.2618\,\delta$	$T(p) = \ln[(1-p)/p]/\ln3$	$G(p) = -\delta[\ln\{p(1-p)\} + \ln2]/\ln3$

Table 7.2. Distributional properties — the skew logistic distribution

parameters. Although historically the two basic forms of the lambda distribution have been developed on a purely empirical basis, we will discuss their construction using the approaches of previous chapters. This will enable further constructions to be considered.

The three-parameter, symmetric, Tukey-lambda distribution

Consider the member of the reflection family of distributions constructed from a power distribution and its reflection. This gives in basic form

$$S(p) = p^\alpha - (1 - p)^\alpha. \ \alpha \geq 0.$$

If we do the same for the Pareto and its reflection we similarly obtain

$$S(p) = (1 - p)^{-\alpha} - p^{-\alpha}. \ \alpha \geq 0.$$

A natural modification is to merge these by dropping the requirement that $\alpha \geq 0$. However, if α is negative in the first formula, then z is not an increasing function of p. This situation is remedied by the adjustment of the scale parameter to give the three-parameter form as

$$Q(p) = \lambda + (\eta/\alpha) \ [p^\alpha - (1 - p)^\alpha]. \ -\infty < \alpha < \infty.$$

Now $Q(p)$ is always an increasing function of p. Notice that there are no explicit CDF or PDF for this distribution. Figure 7.1 illustrates the shapes the distribution can take, presenting it as a special case of a version with two-shape parameters that are equal here. Table 7.3 gives some detail. The distribution is called the **symmetric lambda distribution** or the **Tukey-lambda distribution**. The parameter α is the shape parameter. From the above, it is seen that for positive α the model is based on the power distribution and has a finite distributional range $(\lambda - (\eta/\alpha), \lambda + (\eta/\alpha))$. For negative α it is based on the Pareto and has infinite range $(-\infty, \infty)$. The limiting case as α approaches zero corresponds to the symmetric logistic distribution.

Table 7.3 clearly illustrates the versatility of the symmetric lambda as a potential model for a range of data shapes. The name of the family arises from the conventional use of the Greek lambda for the power in the definition. We have, however, kept to the notation convention of the rest of the book. The main properties of the distribution are tabulated in Table 7.4.

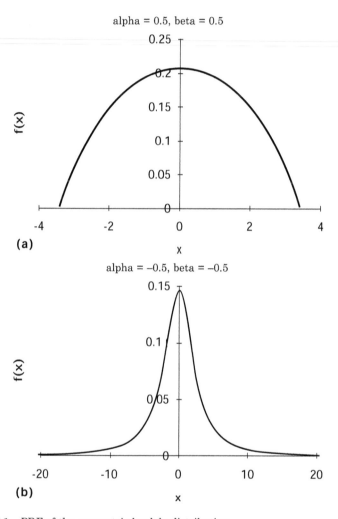

Figure 7.1. PDF of the symmetric lambda distribution

The four-parameter lambda

The simplest way to extend the symmetric lambda to cover skew distributions is to use three linear parameters

$$x_p = \lambda + (\eta/\alpha)[\theta p^\alpha - (1 - p)^\alpha]. \quad -\infty < \alpha, \theta < \infty.$$

Condition on α	Shape	Range of Distribution
$\alpha < 0$	Unimodal, heavy Pareto tails	$-\infty < x < \infty$.
$\alpha \Rightarrow 0$	Logistic distribution	$-\infty < x < \infty$.
$0 < \alpha < 1$	Unimodal	$-1/\alpha \leq x \leq 1/\alpha$
$\alpha \approx 0.135$	Approximately normal	$-7.4 \leq x \leq 7.4$
$\alpha = 1$	Uniform distribution	$-1 \leq x \leq 1$
$1 < \alpha < 2$	Slightly U-shaped. Power tails	$-1/\alpha \leq x \leq 1/\alpha$
$\alpha = 2$	Uniform distribution	$-1/2 \leq x \leq 1/2$
$\alpha > 2$	Peaked distributions, truncated decaying Power tails	$-1/\alpha \leq x \leq 1/\alpha$

Table 7.3. Shapes of the standard symmetric lambda distribution

$S(p)$	$(1/\alpha)[p^\alpha - (1-p)^\alpha]$.
$s(p)$	$[p^{\alpha-1} + (1-p)^{\alpha-1}]$.
$M = 0$	$IQR = 2(3^\alpha - 1)/(\alpha 4^\alpha)$. $QD = 0$
$\mu_1 = 0$	$\mu_2 = (2/\alpha^2)[\{1/(2\alpha + 1)\} - \{(\Gamma(1 + \alpha))^2/\Gamma(2 + 2\alpha)\}]$
	$\mu_4 = (2/\alpha^4)[\{1/(4\alpha + 1)\} - \{4 \cdot \Gamma(1+3\alpha) \cdot \Gamma(1 + \alpha)/\Gamma$
	$(2 + 4\alpha)\} + \{(\Gamma(1 + 2\alpha))^2/\Gamma(2 + 4\alpha)\}]$.

Table 7.4. Distribution properties — symmetric lambda distribution

This was used by Shapiro and Wilk (1965) and investigated by Ramberg (1975). Following our form for skew distributions we reparameterize this as

$$Q(p) = \lambda + \eta\delta/(2^\alpha\alpha) + (\eta/2\alpha)\,[(1 - \delta)p^\alpha - (1 + \delta)(1 - p)^\alpha]$$

where $-\infty < \alpha < \infty$. $-1 < \delta < 1$. Here an adjustment of the position term has been made so that λ is still the median. The properties of the basic form distribution are given in Table 7.5.

The generalized lambda

The generalized lambda is a four-parameter generalisation of the three-parameter model that uses two linear and two non-linear parameters. See Karian and Dudewicz (2000). The original form of the model (Form 1, Ramberg and Schmeiser (1974)) is

$$Q(p) = \lambda + \eta[p^\alpha - (1 - p)^\beta].$$

Using the model in this simple form requires some conditions on α and β to ensure an increasing function of p. An alternative form, Form 2, the canonical form, based on generalizing and centring the power and Pareto forms, ensures a valid quantile function for all parameter values. This was proposed by Freimer et al. (1988) and is written as

$$Q(p) = \lambda + \eta[(p^\alpha - 1)/\alpha - \{(1 - p)^\beta - 1\}/\beta].$$

This form enables the limiting cases where α and β tend to zero to give exponential tails. Table 7.6 indicates the behaviour of the distribution for varying α and β. In part (b) of the table the shapes of the left and right tail are shown for varying α and β. It is clear that the

$S(p)$	$(\eta/2\alpha)[(1 - \delta)p^\alpha - (1 + \delta)(1 - p)^\alpha]$	$\lambda \Rightarrow \lambda + \eta\delta/(2^\alpha\alpha)$
$s(p)$	$(\eta/2)[(1 - \delta)p^{\alpha-1} - (1 + \delta)(1 - p)^{\alpha-1}]$	
$M = 0$	$IQR = \eta(3^\alpha - 1)/(4^\alpha \cdot \alpha)$	
	$QD = \eta\delta(2 \cdot 2^\alpha - 1 - 3^\alpha)/(4^\alpha\alpha)$	
	$G = \delta(2 \cdot 2^\alpha - 3^\alpha - 1)/(3^\alpha - 1)$	$\alpha > 1$ skewness has sign $-\delta$
	$T(p) = 4^\alpha[(1 - p)^\alpha - p^\alpha]/(3^\alpha - 1)$	

Table 7.5. Distributional properties — four-parameter lambda distribution

Special Cases

α	β	Form of distribution
0	0	Logistic
∞	0	Exponential
1	1	Uniform
1	∞	Uniform
∞	1	Uniform
2	2	Uniform

General shapes of the two tails

Term	Condition	Tail Range	Form
Left Tail	$\alpha < 0$	$(-\infty, 0)$	Reflected centred Pareto
$(p^\alpha - 1)/\alpha$	$\alpha = 0$	$(-\infty, 0)$	Reflected exponential
	$0 < \alpha < 1$	$(-1/\alpha, 0)$	Shifted power, decreasing to left
	$\alpha = 1$	$(-1, 0)$	Uniform
	$\alpha > 1$	$(-1/\alpha, 0)$	Shifted power, increasing to left
Right Tail	$\beta < 0$	$(0, \infty)$	Centred Pareto
$-[(1-p)^\beta - 1]/\beta$	$\beta = 0$	$(0, \infty)$	Exponential
	$0 < \beta < 1$	$(0, 1/\beta)$	Shifted reflected power, decreasing to right
	$\beta = 1$	$(0, 1)$	Uniform
	$\beta > 1$	$(0, 1/\beta)$	Shifted reflected power, increasing to right

Table 7.6. Shapes and properties of the canonical generalized lambda

generalized lambda is built on the basis of the addition rule to give a rich variety of distributional shapes. It is also evident that the parameters α and β determine not only the shape but also the relative weights of the tails. One consequence of this is that skewness is modelled as a result of tail shape and not as an independent feature. A further consequence is that when both parameters tend to zero, the model tends to the symmetric logistic and not the skew logistic. Figure 7.2 illustrates some shapes for Form 2.

A paper by Freimer, Mudholkar, Kollia and Lin (1988) analyzes the shapes of the distribution and classifies five shapes as in Table 7.7.

Notice that Form 1 and Form 2 differ in position and scale parameters but not in terms of basic shape. Form 2 is needed when it is not known what form the data may take and to ensure a meaningful set of parameters on estimation. It is often the case in practice that past knowledge of the situation implies knowledge of the sign and rough range of α and β. In these cases there is no need for the additional

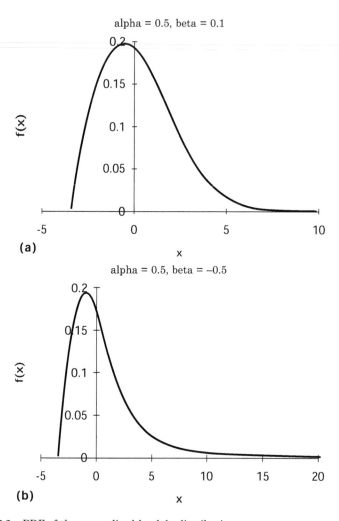

Figure 7.2. PDF of the generalized lambda distribution

Class	α	β	Distributional form
I	<1	<1	Unimodal
II	>1	<1	Monotone
III	[1, 2]	[1, 2]	U-shaped
IV	>2	[1, 2]	S-shaped
V	>2	>2	Unimodal

Table 7.7. Classes of generalized lambda distributions

complexity of Form 2. Using the standard version of Form 1 the main quantile properties flow simply from the formula. (For Form 2 the modifications are simple and fairly obvious.) The results are

$$M = 1/2^\alpha - 1/2^\beta. \; LQ = 1/4^\alpha - 3^\beta/4^\beta. \; UQ = 3^\alpha/4^\alpha - 1/4^\beta.$$

$$IQR = (3^\alpha - 1)/4^\alpha + (3^\beta - 1)/4^\beta.$$

$$QD = (3^\alpha + 1)/4^\alpha - (3^\beta + 1)/4^\beta - 1/2^{\alpha-1} + 1/2^{\beta-1}$$

The moment properties are obtained by the use of the formulae given in Section 4.6. The expectation and variance are

$$\mu_1 = \lambda + \eta[1/(\alpha + 1) - 1/(\beta + 1)]$$

$$\mu_2 = \eta^2[\{1/(2\alpha + 1) - 2B(\alpha + 1, \beta + 1) + 1/(2\beta + 1)\} \\ - \{1/(\alpha + 1) - 1/(\beta + 1)\}^2]$$

where $B(\,,\,)$ is the beta function (see Appendix 1).

The five-parameter lambda

One criticism of the generalized lambda referred to above is that the shape parameters α and β also determine the skewness. It seems reasonable that there should be three linear parameters determining position, scale, and skewness and two parameters determining the shapes of the two tails. This suggests a natural generalisation of the four-parameter model to give a **five-parameter lambda distribution**. This is defined for Form 1 by

$$Q(p) = \lambda + (\eta/2)[(1 - \delta)p^\alpha - (1 + \delta)(1 - p)^\beta]. \; -\infty < \alpha < \infty. \; -1 \le \delta \le 1.$$

For Form 2 we have

$$Q(p) = \lambda + (\eta/2)[(1 - \delta)(p^\alpha - 1)/\alpha - (1 + \delta)\{(1 - p)^\beta - 1\}/\beta].$$

When the shape parameters tend to zero this becomes the skew logistic. If the Galton skewness coefficient is derived for this model it becomes evident that the skewness is determined by both the shape parameters and the skewness parameter. Thus there needs to be some way of forcing the shape parameters to control shape without affecting the relative weight of the two distributions. A similar point relates to the position parameter, since the median depends on all the parameters. A partial solution to these problems is to standardize the two tail distributions so that they have medians at −1 and +1, respectively, and to adjust the constant term so that the overall median lies at λ. This generates Form 3:

$$Q(p) = \lambda - \eta\delta + (\eta/2)[(1 - \delta)(p^\alpha - 1)/(1 - 0.5^\alpha) \\ + (1 + \delta)\{(1 - p)^\beta - 1\}/(0.5^\beta - 1)].$$

Although the shape parameters still influence the skewness, the constraint means that the skewness parameter does directly control the weight given to two similar tails.

It may be noted that the five-parameter form was referred to by Joiner and Rosenblatt (1971). Figure 7.3 illustrates the shape of the distribution.

7.4   Extreme value distributions

In Section 3.2 it was proved that the distribution of the largest and smallest observations of a sample of n, the **extremes**, have particularly simple quantile functions, for example, that of the largest observation was $Q(p^{1/n})$. Clearly, each distribution will have its own extreme distributions. There are many situations where extremes are of interest. For example, distributions of maximum flood heights, wind velocities, temperatures, wave heights, minimum rainfall, and lifetimes. An inherent aspect of these situation is that data tends to be continually collected, so that n is increasing. Also, our interest is often in maximum values for increasing periods, e.g., the expected maximum flood heights to be experienced in 10, 50, 100, 200, etc. years. As a consequence of this, there is interest in the changes in the distribution of the extremes as n increases. It can be proved, originally by Fisher and Tippett (1928) (see, for example, Arnold, Balakrishnan and Nagaraja (1992)), that as n tends to infinity the distributions of the extremes tend to one of three limiting distributions; which one depends on the form of the distribution $Q(p)$. The most commonly occurring of these limiting distributions has standard quantile function

$$S(p) = -\ln\,[-\ln(p)], \text{ with distributional range } (-\infty, \infty).$$

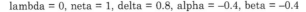

lambda = 0, neta = 1, delta = 0.8, alpha = –0.4, beta = –0.4

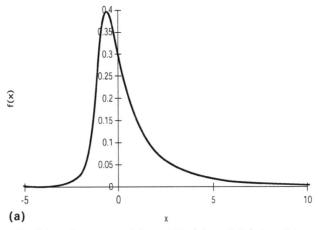

(a)

lambda = 0, neta = 1, delta = 0.5, alpha = 0.4, beta = 0.4

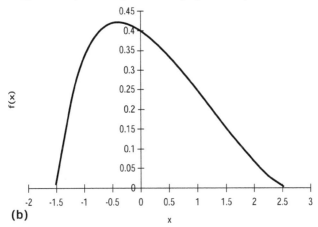

(b)

Figure 7.3. PDF of the five-parameter lambda distribution

This distribution is called the **type 1 extreme value (EV) distri-bution,** or sometimes the Gumbel distribution, and its main properties were given in Section 5.8. This distribution has been extensively used in the study of the types of situations indicated above and we will devote a brief section later to this area of application. It should be noted that we have defined the distribution so as to have positive skewness. The reflected version of this model, with negative skewness but the same doubly infinite range, is also sometimes called the extreme value distribution with its reflection called the reversed EV distribution, so some care needs to be taken when referring to the literature. See Johnson, Kotz and Balakrishnan (1995, Volume 2) for a survey of the literature.

Any distribution whose tails tend to die away exponentially has the EV distribution as the limiting form for its extreme observations. For other distributions, the limiting forms have quantile functions that are identical save for the sign of one parameter. These are

$S(p) = (-\ln p)^{-\beta}$, DR $(0, \infty)$, The type 2 Extreme Value Distribution;

$S(p) = -(-\ln p)^{\beta}$, DR $(-\infty, 0)$, The type 3 Extreme Value Distribution.

There are just three forms for the limiting distributions. The terminology refers to the **domain of attraction** of a given type of EV distribution as being those distributions whose extreme distributions converge to the given distribution as n increases. A majority of standard distributions are within the domain of attraction of the type 1 EV distribution.

If we again use the generalizing technique of previous sections, we can write all three models in canonical form as

$$z_p = [1 - (-\ln p)^{\beta}]/\beta, \quad -\infty < \beta < \infty.$$

This form covers all three distributions. The generalized model is called the **generalized extreme value distribution, GEV**. All three distributions are related to the Weibull distribution by Q-transformations (Figure 7.4 illustrates).

The major quantile properties for the standard GEV are from simple substitution:

$$M = [1 - (\ln 2)^{\beta}]/\beta, \quad IQR = [(\ln 4)^{\beta} - \{\ln(4/3)\}^{\beta}]/\beta$$

$$QD = [2(\ln 2)^{\beta} - (\ln 4)^{\beta} - \{\ln(4/3)\}^{\beta}]/\beta.$$

There have been many studies of the type 1 EV distribution, which is by far the most common in theory and has been in use since the 1930s. Although it often provides a good approximation to many practical sets of data, there are often situations that would require many millions of observations for the limiting distribution to closely approach the type 1 limiting distribution. Interest has, therefore, focused recently on a so-called penultimate approximation that gives a reasonably close fit to the tail of interest for large but real sets of

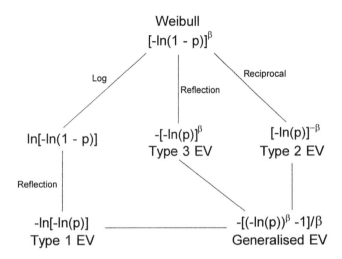

Figure 7.4. Transformational links for the extreme value distributions

data. A distribution that provides a good approximation is in fact the GEV, which conveniently includes the type 1 EV as a special case. A consequence of this is that for practical analysis and application the GEV is now commonly used.

7.5   The Burr family of distributions

In 1942 Irving Burr introduced a distribution with CDF:

$$p = F(x) = 1 - (1 + x^{1/\beta})^{-1/\alpha}, \quad \alpha, \beta > 0.$$

Re-expressing this in quantile form gives the quantile function

$$S(p) = [-1 + 1/(1 - p)^\alpha]^\beta.$$

This is the Pareto distribution centred to the distributional range $(0, \infty)$ and then transformed by the power Q-transformation. In present terminology this is the Burr XII distribution. Burr extended this model by observing that the reciprocal transformation extends the forms of distributional shape covered. To achieve this with quantiles we use the reciprocal rule. This gives the Burr III distribution (also sometimes called the kappa distribution):

$$S(p) = [p^{\alpha}/(1 - p^{\alpha})]^{\beta}, \text{ with DR } 0 < x < \infty.$$

This is the power transformation of the variable with quantile function:

$$S(p) = p^{\alpha}/(1 - p^{\alpha}), \text{ with } 0 \le x < \infty.$$

If we take a straight logarithmic transformation of this we obtain

$$S(p) = \ln[p^{\alpha}/(1 - p^{\alpha})], \text{ with } -\infty < x < \infty.$$

This the Burr II distribution. This is also sometimes called the generalized logistic distribution and could also be called the exponentiated logistic, following the language of the exponentiated Weibull. (A series of papers by Burr are given in the references.) Notice that the Burr II and Burr III can be merged to canonical form, using the generalization approach of a suitable change of scale and position giving a common expression

$$S(p) = [\{p^{\alpha}/(1 - p^{\alpha})\}^{\beta} - 1]/\beta. \text{ with } -1/\beta \le x < \infty.$$

The limiting case of $\beta = 0$ then gives the Burr II from the Burr III form.

All the members of the Burr family are given by solutions of a general form of differential equation. The majority do not have explicit quantile form. There are, however, two more, less common forms that do. They are

The Burr IV $S(p) = \beta/[1 + \{(1 - p^{\alpha})/p^{\alpha}\}^{\beta}]$, with $0 \le x \le \beta$.

The Burr X $S(p) = [-\ln(1 - p^{\alpha})]^{0.5}$, with $0 \le x < \infty$.

This last is a special case of the exponentiated Weibull distribution. Figure 7.5 shows the transformational links between some of the members of the Burr family.

7.6   Sampling distributions

Quantile functions do not readily lend themselves to the analysis of statistics formed by calculations with sample data. For example, they do not help in finding the distributions of sums or sums of squares of

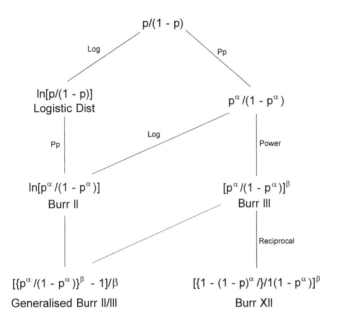

Figure 7.5. Transformational links for some of the Burr distributions

observations. The theory of these relies mainly on probability density functions. Nonetheless when one wishes to use these distributions for the analysis of data their quantile functions are needed. In the pre-computer era most books of statistical tables included a number of tables of quantile values for specified ranges of p values for the common sampling distributions. Now it is expected that the computer software will derive the values of $Q(p)$ for any desired values of p. All statistical software and probably most spreadsheets make at least some quantile functions available. Table 7.8 gives a list of the most commonly used quantile functions with their designations in SAS™ (the registered trademark of the SAS Institute Inc.) and Excel™. The table also gives the expressions for some other distributions of common use that lack explicit quantile functions.

7.7   Discrete distributions

Introduction

Our studies to this point have focused on the distributions of continuous variables. The aim of this section is just to draw attention to the

Distribution	SAS™ Designation	Excel™ Designation
Normal		
standard	PROBIT(p)	NORMSINV(p)
non-standard		NORMINV(p, μ, σ)
Log-Normal		LOGINV(p, μ, σ)
t-distribution	TINV($p,df,nc>$)	TINV(p, df)
Chi-Squared	CINV(p,df,nc)	CHIINV($1 - p$, df)
Gamma	GAMINV(p, α)	GAMMAINV(p, α)
F-distribution	FINV(p, $df1$, $df2$, nc)	FINV(p, $df1$, $df2$)
Beta Distribution	BETAINV(p, α, β)	BETAINV(p, α, β)

Note: df = degrees of freedom, Greek symbols = parameters, nc = numeric noncentrality parameter ≥ 0.

Table 7.8. Commonly available quantile functions

fact that it is sometimes possible to develop straightforward quantile functions for discrete variables. The forms of CDF and QF for discrete distributions are step functions, as was illustrated in Figure 2.1. As these are not continuous functions, neither has a simple inverse. In general, discrete variables have CDF that involve elaborate summations that preclude any direct inversion. There are, however, some exceptions. We illustrate with two examples.

The geometric distribution

Suppose a series of trials are made, continuing until a specific result is obtained at trial r. For example, I keep repeating a complex task until I succeed. If θ is the probability of succeeding and this remains constant (there is no improvement with practice), then the probability, assuming independent trials, of the random variable R being r is, by the multiplication law of probability, the probability of $r - 1$ failures times the success probability.

$$Pr(R = r) = (1 - \theta)^{r-1} \theta. \ r = 1, 2, 3, ...$$

Hence the CDF is

$$p = F(r) = \Sigma_1^r (1-\theta)^{l-1} \theta.$$

If we use the result that, if $0 < a < 1$, the series $1, a, a^2, a^3, ..., a^m$ sums to $(1 - a^{m+1})/(1 - a)$ then

$$p = \theta[1 - (1 - \theta)^r]/[1 - (1 - \theta)]$$
$$= 1 - (1 - \theta)^r$$

Hence

$$\ln(1 - p) = \ln[(1 - \theta)^r]$$

and

$$r = \ln(1 - p)/\ln(1 - \theta).$$

Care needs to be taken with this expression since r cannot take all the continuous values implied by this formula, for it is limited to the integers 1, 2, 3, ... In fact, we need to round the value up to the nearest integer. The simplest way to do this is to add one to the integer part of the value given by the formula. Thus the appropriate quantile function is

$$S(p) = INT[\ln(1 - p)/\ln (1 - \theta)] + 1.$$

where INT[ ] is the function, available in most relevant software, that takes the integer part of a value.

The binomial distribution

There are some situations where the previous calculations cannot be carried out but nonetheless a quantile function is obtainable. The binomial distribution models the same situation as the geometric model with independent trials of constant probability. However, in this case r is the number of "successes" in a fixed number of trials, n. Here there is no direct method for obtaining the quantile function. However, as this is a much used distribution, statistical software and spread-sheets usually give a suitable function. This is not usually expressed as a quantile function but rather in terms of the value needed to carry out a particular statistical test. However, this is in fact a quantile function. In Excel™, for example, the function CRITBINOM(n, θ, p) gives the required quantile function.

7.8   Problems

1. For the four-parameter lambda distribution show that for negative α the tails stretch out to infinity. For positive α, show that the distributional range is obtained from

$$Q(0) = \lambda + \eta\delta/(2^\alpha\alpha) - (\eta/2\alpha)(1 + \delta)$$

$$Q(1) = \lambda + \eta\delta/(2^\alpha\alpha) + (\eta/2\alpha)(1 - \delta).$$

There is thus a finite distributional range of width η/α. Show also that there is a smooth approach to these finite limits for $0 < \alpha < 1$. For $\alpha > 1$ show that there is a visible cutoff at the ends of the distribution.

2. For the generalized lambda distribution on page 160, derive the values of $f_p(p)$ and $df_p(p)/dp$ and hence the end shapes of the distribution for varying α and β.

3. Explore and plot the properties of the generalized Burr II/III given by

$$S(p) = [\{p^\alpha/(1 - p^\alpha)\}^\beta - 1]/\beta, \text{ with } -1/\beta \leq x < \infty.$$

4. The most common sampling distribution is the normal, since the sum of normal observations is itself normal and many distributions tend to normality. Show that if $N(p)$ is the normal quantile function, then $N((1 + p)/2)$ gives the **half-normal distribution**, which has the form of the right-hand tail of the normal. Hence, examine the shapes of the two variants of the skew normal distribution given by

$$N_1(p) = \lambda + (\eta/2)[(1 + \delta)N((1 + p)/2) - (1 - \delta)N((2 - p)/2)].$$

$$N_2(p) = \lambda + \eta[(1 + \delta)\{p^2 (3 - 2p)\}N((1 + p)/2) - (1 - \delta)$$
$$\{1 - p^2 (3 - 2p)\}N((2 - p)/2)].$$

CHAPTER 8

Identification

8.1 Introduction

In the previous chapters we have looked at a range of methods of building models of distributions. In this chapter we look at the identification of suitable models for a given set of data. There are a number of aspects of identification. There is an initial need to understand the context in which the data arose, then the data needs to be pictured in as many ways as possible. This is done mainly by using graphical techniques, many of which we have already discussed. We may need some numerical information about the data. On the basis of these studies we may have ruled out some possibilities and highlighted others. The next stage is to do a range of comparisons with a list of candidate models. This may involve separate studies of the tails of the data. From this stage we will hopefully home onto a few real candidates. These will need to be compared with the data in detail. It is advisable at this last identification stage to select more than one model. Having obtained "the chosen few," these are fitted to the data, using the methods in Chapter 9. The methods of validation, discussed in Chapter 10, are then used to finally decide on the model.

8.2 Exploring the data

The context

When starting the identification stage of modelling it is important to use all that is known about the context and source of the data. Questions should be asked about the underlying mechanism generating the data and any previous relevant studies. Some of the information so obtained may lead directly towards a choice of model. Relevant questions are

(a) Is anything known of the distributional range, e.g., are the variables inherently positive? Is there a natural lower threshold or perhaps an upper bound?

(b) Are the variables discrete or continuous? Our focus is on continuous distributions; nonetheless, we can often use such distributions for essentially discrete variables, e.g., financial data is inherently discrete but usually treated as continuous. If this is the case, we may need to be aware of this underlying discreteness.

(c) Is there anything known that will indicate shape features of the distribution? For example, is there an inherent symmetry or a natural skewness in a known direction?

(d) Is anything known from previous studies that might be helpful? For example, if similar data has been collected before, there will be some knowledge of the practical limits to the data, which will help in the consideration of possible outliers. There may also be knowledge of particular features or success of different approaches to analysis. It might even be known, for example, that the Weibull distribution gave a good fit to previous similar data. Most applied statistical situations are not unique. There is a history that will probably throw light on the new data and situation.

We have already discussed in Chapters 1 and 2 a variety of numerical and graphical means of studying a set of raw data. Although in this chapter some further approaches will be described, it is convenient to summarize the set of graphical tools using tables and illustrations.

Numerical summaries

An initial set of numerical summaries are given by the five-number summary, *iqr* and *g*. The quartiles may not correspond to data points. We therefore need to use the more formal calculation for a quantile based on the formula given in Section 2.2. Thus for the quartiles we have

$$lq = (1 - h)x_{[r]} + hx_{[r] + 1}, \ uq = (1 - h)x_{(n + 1 - [r])} + hx_{(n - [r])}.$$

where $r = n/4 + 0.5$ and $h = r - [r]$.

General shape

Using the methods given in Chapters 1 and 2, the general shape of
the distribution can be explored with the plots given in Table 8.1.

Sample Function	Of	Against	Comments
Quantile, $\tilde{Q}(p)$	$x_{(r)}$	p_r	
CDF, $\tilde{F}(x)$	p_r	$x_{(r)}$	
Quantile density,	Dx/Dp	mid-p_r	Raw and smoothed plots,
$\tilde{q}(p)$			see Sections 1.3 and 4.9
p-PDF, $\tilde{f}_p(p)$	Dp/Dx	mid-p_r	Raw and smoothed plots
PDF, $\tilde{f}(x)$	Dp/Dx	mid-$x_{(r)}$	Raw and smoothed plots

Table 8.1. General shape plots

Four important quantities that can usually be visually estimated
from these plots are the limiting values for the distributional range
$(\tilde{Q}(0), \tilde{Q}(1))$ and the density function $\tilde{f}_p(0)$, $\tilde{f}_p(1)$. The value of $\tilde{Q}(0)$
is particularly important. Knowing whether the distribution has the
distributional range $(-\infty, \infty)$ or $(0, \infty)$ is critical in guiding an initial
selection of potential models; however, the answer may not be clear
from the data, even if it is all positive. For example, many situations
with positive data still use the normal distribution. The justification
is that the probability of negative observations arising from the fitted
model is made infinitesimal by the choice of mean and standard
deviation. Thus there is a certain amount of trial and error in such
choices. A combination of empirical and conceptual arguments may
be needed to decide whether to use a model on the positive axis with
a position parameter of zero, i.e., a threshold at zero, or a two-tailed
distribution positioned well to the right of the origin. Where a threshold is needed this can usually be simply shifted to give a threshold
at $\lambda = 0$. In such cases the scale parameter becomes the main con-
trolling parameter.

Skewness

Table 8.2 gives a number of plots that indicate skewness in a set of
data. Interest initially is in the simple fact of skewness or symmetry
in the data. At a later stage these plots may be referred to in relation
to the form of the skewness.

Plot	Of	Against	Comments
Deviations	$x_{(n+1-r)} - m$	$m - x_{(r)}$	Include 45° line for symmetry
$g(p)$	$[x_{(n+1-r)} + x_{(r)} - 2m]/iqr$	p_r	
$g^*(p)$	$[x_{(n+1-r)} + x_{(r)} - 2m]$ $/[x_{(n+1-r)} - x_{(r)}]$	p_r	
Folded CDF	$\tilde{F}(x_{(r)})$	$x_{(r)}$ for $p_{(r)} \le 0.5$	See Monti
[Mountain plot]	$1 - \tilde{F}(x_{(r)})$	$x_{(r)}$ for $p_{(r)} > 0.5$	(1995)
Spacing plot	$x_{(n+1-r)} - x_{(n-r)}$	$x_{(r+1)} - x_{(r)}$	Include 45° line for symmetry

Table 8.2. Plots for indicating skewness

Tail shape

Table 8.3 gives a number of plots that may be used to look at the tail shapes. There are two groups of these. One uses all the data, the other looks separately at each tail.

Plot	Of	Against	Comments
Shape index, $t(p)$	$[x_{(n+1-r)} - x_{(r)}]/iqr$	p_r	$0 < p_r < 0.5$
Spacing plot	$x_{(n+1-r)} - x_{(n-r)}$	$x_{(r+1)} - x_{(r)}$	
Upper shape index, $ut(p)$	$[x_{(n+1-r)} - m]/[uq - m]$	$1 - p_r$	$0 < p_r < 0.5$
Lower shape index, $lt(p)$	$[m - x_{(r)}]/[m - lq]$	p_r	$0 < p_r < 0.5$
$q(p)/Q(p)$	$2(x_{(r+1)} - x_{(r)})$ $(n+1)/(x_{(r+1)} + x_{(r)}).$	p_r	

Table 8.3. Tail-shape plots

Interpretation

No attempt has been made to discuss the detailed interpretation of these many plots as it would be too space consuming. The recommended approach is to

(a) Construct a set of the plots for the data.
(b) Develop a small number of potential distributions, for the whole model or from each tail.

(c)   For the chosen or constructed models, create profiles (see Section 4.4) of size n, where n is the size of the sample being studied and some sensible parameter values are assumed.

(d)   Re-plot the set of exploratory graphics using the profile data for each of the distributions.

(e)   Compare the results of (a) and (d) and modify the models if required.

Where needed, the data might be standardized. Suitable parameters can be chosen using the ability of most computer graphics to dynamically adjust the plots as the parameters are altered. This dynamic graphics facility enables the type of explorations needed for identification to be carried out with relative ease.

8.3   Selecting the models

Starting points

Having obtained a good feel of the data, it is now necessary to select the models to use. A classic approach at this stage has been to choose just one model. Unfortunately this tends to hide the often very subjective element in the choice. Once chosen, the model becomes a hidden assumption in the use of statistical methods. Sophisticated statistics can be used, estimates can be quoted to four significant figures, and predictions made, all on the basis of an assumed model that was only marginally better than some very different model. It seems much more cautious to choose at least two models and carry through the estimation on both, then choose which to finally use at the validation stage, which will also take note of the specific application of the models. The final choice may indeed depend on non-statistical considerations.

One consideration in identification relates to what is called the **principle of parsimony**. This is essentially a principle of simplicity that recommends using a simple model in preference to a more complex model. In our context this simplicity may be taken as, for example, preferring to use a three-parameter model instead of a four-parameter model, or a linear model rather than a non-linear model. One element of the logic underlying this principle is that as we build models we tend to find ourselves involved in an iterative, improvement process. We compare a better set of data with our originally identified model and find that the model can be developed further. In general, it is clearer to compare new data using simple models than complex ones. The systematic deviations stand out more clearly against a simple plot, e.g., against a straight line, rather than a more complex one. The overriding consideration in modelling has to be how well the model performs in its application; the principle of parsimony provides a useful general guide.

In our current context there are two different model types: those modelled by some basic model and those that require a model to be built up, for example, by addition of right and left quantile functions. One outcome of the studies of Section 8.1 should be a feel for which of these is appropriate. The study should have revealed the likely values of $Q(0)$, $Q(1)$, $f_p(0)$ and $f_p(1)$. These put considerable constraints on the possible distributions, as is shown in Table 8.4. We start our study of model choice on the assumption that the data comes from a single basic model.

Identification plots

The basic approach to model-based identification is the quantile plot of $x_{(r)}$ against some model $\lambda + \eta S(p_r)$. The position and shape parameters alter the intercept and slope but not the straightness of the line for the correct model. Thus we can ignore these two parameters in most of the identification considerations and concentrate on $S(p)$. Where it is helpful, the data may be standardized by subtracting the median or threshold and/or dividing by the interquartile range.

$S(0)$	$S(1)$	$f_p(0)$	$f_p(1)$	Distributions
$-\infty$	∞	0	0	Logistic, skew
				Logistic, normal, t,
				Extreme value (1)
				Lambda family[a]
0	∞	0	0	Weibull[a]
				Gamma, χ^2
				log Normal
0	∞	1	0	Exponential
0	∞	∞	0	Centred Pareto, Weibull[a]
0	1	0 or ∞	c	Power[a]

Note: Positive constant c. [a] = applies to specific ranges of the shape parameter.

Table 8.4. Tail values for some common distributions

Interpreting Q-Q plots

A good choice of $S(p)$ will lead to a roughly linear Q-Q plot. If the plot veers away at the ends, it is indicating that the distribution needs to have either shorter or longer tails at that end. Figure 8.1 indicates the situations and their interpretations for the distribution. It may be that the overall shape of the plot corresponds to that of a simple function $T(.)$. In this case the transformation rule gives the suggested distribution as $T(Q(p))$, which can then be used for the next Q-Q plot.

Fowlkes (1987) gives Q-Q plots of a wide variety of distributions against each other. The plots for the data against different models may match up with those of a given model against the same set of distributions, thus identifying the model. As there are hundreds of plots that theoretically could be compared, this method in practice requires an initial narrowing of the list of potential options, underlining again the need to use all available background information to develop short lists of possible models.

Identification plots for common distributions

Section 6.3 presented a set of Q-transformations from the uniform distribution as a means of generating a wide variety of distributions, all in standard form, $S(p)$. These provide a set of standard models with which we can compare the data in situations where we believe that a single model is appropriate. Let us start by considering only situations where the data exploration has indicated a distribution with $Q(0) = 0$

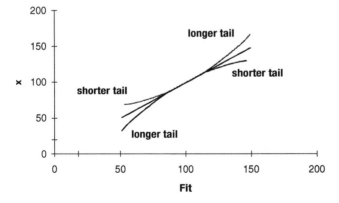

Figure 8.1. Interpreting fit-observation diagrams

$R(p)$	Plot	Against	Intercept	Slope
Power	$\ln x$	$\ln p$	$\ln \eta$	β
Pareto	$\ln x$	$-\ln q$	$\ln \eta$	β
Exponential	x	$-\ln q$	0	η
Burr III	$\ln x$	$\ln(p/q)$	$\ln \eta$	β
Weibull	$\ln x$	$\ln(-\ln q)$	$\ln \eta$	β
Extreme Value Type 2	$\ln x$	$-\ln(-\ln p)$	$\ln \eta$	β

Note: Distributions of the form $\eta R(p)$ or $\eta R(p{:}\beta)$; $q = 1 - p$; X in the range $(0, +)$.

Table 8.5. Identification plots for positive distributions

or some constant that can be removed by adjusting the data. Hence the underlying model is of the form $x = \eta R(p)$, where $R(p)$ is a right-tailed distribution on $(0, -)$. Suppose, for example, we had a set of data that came from an exponential distribution, which could be represented as LVp. If we denote $1 - p_r = q_r$ and observe that $LVp = -Lq$, then a plot of $x_{(r)}$ against $-\ln(q_r)$ will be a straight line of slope η, which is unknown but nonetheless a constant. The Weibull distribution is simply an exponential raised to a power, $PLVp$. If we take logs of the observations, the power becomes a multiplier of the logarithm and a plot of $\ln(z_r)$ against $\ln(-\ln q_r)$ is again a straight line, with the slope equal to the power in the Weibull formula and the intercept depending on η. Table 8.5 gives the plotting values of x and of p and q that generate straight lines in this way. It will be seen that the plots of x and $\ln(x)$ against $\ln(p)$, $-\ln(q)$, etc. cover most of the standard distributions on the positive axis of x. Figure 8.2 illustrates the plots for a set of data and the relative linearity of some of the plots is clear.

It will be seen that selecting on the basis of linearity in plots is a highly convenient approach. Some care needs to be taken as the differences in scales mean that specific deviations are not directly comparable. Notice that the intercepts and slopes of the lines give rough estimates of the model parameters. However, interest here lies simply in the linearity of the plots as a means of initial identification.

It may be noted that plotting the observations against simple functions of p gives the Q-Q plots for some of the distributions with distributional range $(-\infty, \infty)$. These are shown in Table 8.6.

Section 8.1 listed a wide range of plotting variables. If we now have a proposed model these can be plotted against the corresponding quantities for the population. The layout for deriving such plots is standard and is illustrated for the shape index in Table 8.7. If the model is reasonable, then the plot will be approximately linear.

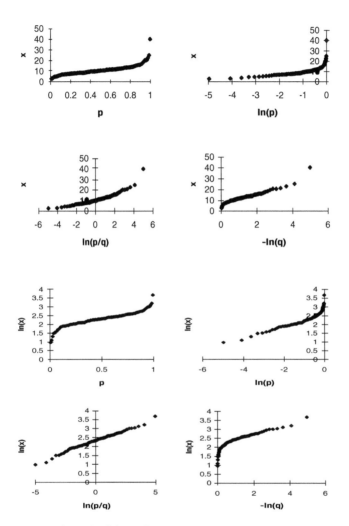

Figure 8.2. A set of standard data plots

$S(p)$	Plot	Against	Intercept	Slope
Logistic	z	$\ln(p/q)$	0	η
Extreme Value Type 1	z	$-\ln(-\ln p)$	0	η
Cauchy	z	$\tan[\pi(p - 0.5)]$	0	η

Note: Distributions of the form $\lambda + \eta S(p)$; $q = 1 - p$; $z = x - \lambda$.

Table 8.6. Identification plots for distributions on $(-\infty, \infty)$

r	p^*	$\beta =$ $Q(p^*)$	x	$iqr =$ $t(p^*)$	$IQR =$ $T(p^*)$
1	p_1^*	$Q(p_1^*)$	$x_{(1)}$	$(x(n) - x(1))/iqr$	$(Q(p_n^*) - Q(p_1^*))/IQR$
2	p_2^*	$Q(p_2^*)$	$x_{(2)}$	$(x(n-1) - x(2))/iqr$	$(Q(p_{(n-1)}^*) - Q(p_2^*))/IQR$
$m = (n+1)/2$	p_m^*	$Q(p_m^*)$	$x_{(m)}$	$(x_{(m+1)} - x_{(m)})/iqr$	$(Q(p_{(m+1)}^*) - Q(p_m^*))/IQR$

Note: $p^* =$ Median-p; $Q(p^*) = Q(p^*;\beta)$; $T(p^*) = T(p^*;\beta)$.

Table 8.7. Layout for sample-population comparisons — example

Straightening plots

We have avoided thus far the issue of plots whose form is influenced by shape parameters. For example, if the data is truly a Tukey lambda and we construct a Q-Q plot based on a Tukey lambda, but with the wrong shape parameter, then we will not get a linear plot. A simple way of avoiding this is to choose any parameters to make the plot as linear as possible. This can be done by using the standard measure of linearity, which is the correlation coefficient defined in Section 4.7. For example, with Q-Q plots the correlation between the ordered observations and their median rankits will give a measure of linearity. As the aim is to find a model whose Q-Q plot is linear, this provides a useful numerical measure of success. The simplest measure first requires the ordered data and median rankits be standardized by subtracting their respective averages, giving $x'_{(r)}$ and M'_r. We then calculate

$$r = \Sigma x'_{(r)} M'_r / \sqrt{[\Sigma x'^2_{(r)} \Sigma M'^2_r]}$$

The correlation coefficient r can be used to compare, in general terms, alternative models used in Q-Q plots. The closer to one the better the model. In this case the alternatives are specified by the shape (and/or skewness) parameters. The correlation is independent of any position or scale parameters. The procedure is thus to get the "best" model by choosing the shape parameter(s) to maximize the correlation, i.e., to straighten the points on the Q-Q plot. Table 8.8 shows the layout of a comparison of three models for a set of data. In each case the parameters are chosen to maximize the correlation. It will be seen that two models perform equally well, although one has fewer parameters.

Using p-transformations

In Section 8.9.1 we considered plots for models based on Q-transformations. Those also involving p-transformations pose a more difficult problem, since all the common ones involve a power transformation with an unknown parameter, which for this section we denote by γ. If we knew γ, we could simply plot on the bottom axis the transformations of Table 8.5 with p^γ and $1 - p^\gamma$ replacing p and q. The correlation provides a simple tool for looking for appropriate values of γ. Using an optimizer, the best values of the parameter can be chosen to give the maximum correlation. Thus a mechanized search can be undertaken before the visual comparisons of the best lines.

				$\alpha =$	0.635	
				$\beta =$	0.287	0.806
		correlation		0.988	0.992	0.991
				Exponential	$Po \times Pa$	Weibull
r	p	x		$-\ln(1 - p)$	$p^{\alpha}/(1 - p)^{\beta}$	$(-\ln(1 - p))^{\beta}$
1	0.043	17.88		0.044	0.138	0.081
2	0.087	28.92		0.091	0.217	0.145
3	0.130	33.00		0.140	0.285	0.205
4	0.174	41.52		0.191	0.348	0.264
5	0.217	42.12		0.245	0.407	0.322
6	0.261	45.60		0.302	0.464	0.381
				etc.		

Table 8.8. Model selection by correlation

Plots for data in frequency tables

Data in frequency table form is only a summary of the raw data and is therefore best avoided; however, if it is the only data available, then the plotting techniques need modifying. The essential modification is that the n pairs of values $(x_{(r)}, p_r)$ are replaced by $m - 1$ values (x_i, p_i), where x_i is the upper boundary of a class interval in the table, p_i is the total proportion of the observations less than x_i, and m is the number of non-zero classes. The last class is not normally used as $p_m = 1$, which may send $Q(p)$ to infinity.

8.4   Identification involving component models

If there are too many options when faced with single distributions, the problem is compounded when one considers a constructed model produced, for example, by a weighted addition of two quantile functions; however, we can make use of **dynamic plotting**. Figure 8.3 shows an example. This makes use of several ideas:

(a) We can create models where the shape depends on differing parameters to dominate the two tail shapes.

(b) Weighting of two quantile functions can be used to create skewness or dominance in the tails.

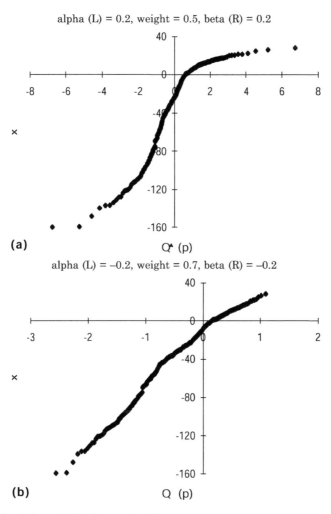

alpha (L) = 0.2, weight = 0.5, beta (R) = 0.2

(a)   Q* (p)

alpha (L) = –0.2, weight = 0.7, beta (R) = –0.2

(b)   Q (p)

Figure 8.3. A dynamic fit-observation diagram

    (c)   Generalized forms can cover a complete range of shape
         parameter values.
    (d)   Computer software creates graphics very rapidly.

Putting these ideas together, Figure 8.3 shows a **Q-Q** plot of a set
of data against a basic five-parameter lambda, ignoring position and
scale. Generalized Pareto distributions form the two tails and a
weighting combines them. The parameters for the left and right tails

and the relative weight are set immediately above the plot. By altering the parameters, the shapes resulting can be explored in a dynamic fashion. In Figure 8.3 positive shape parameters have led to a Q-Q plot showing that the tails of the fitted Paretos are much longer than those of the data. This dynamic approach can also be used in the developing comparisons between profile statistics and sample statistics, such as $T(p)$ and $t(p)$. The layout to obtain suitable graphics was shown in Table 8.7.

8.5   Sequential model building

A frequently used approach to modelling with deterministic construction kits is called sequential model building. This involves a number of stages:

1. A list of model components is prepared and put in some order of potential appropriateness.
2. Some criterion of quality of fit is chosen, perhaps the correlation between the fitted model and data.
3. A component is selected from the list and added to the model (at the start this is the only one).
4. The criterion is evaluated and if reasonably improved the new component is kept; otherwise it is dropped.
5. Return to 3.

The above procedure is called **forward selection**. The procedure is highly developed for deterministic models and available on spreadsheets and statistical programmes. However, there is nothing in the above steps that cannot be applied equally to building a distributional model with quantile function components. The detailed procedure does, however, need a little more consideration. Suppose the first model tried takes the form $Q_1(p) = \lambda + \eta S1(p)$. To simplify later calculations the position and scale parameters are adjusted to correspond to, say, the population median and interquartile range, so that they can be regarded as unchanging during the calculations. In Table 8.9 we try a normal on the data and fit it by least absolutes. We also keep an eye on the correlation between fitted model and data. Provided the correlation is sufficiently large to justify our choice of $Q_1(p)$, we now have a suitable fitted $\hat{Q}_1(p)$. Defining

Initial model — Normal, $\hat{\lambda} = 6.330$, $\hat{\eta} = 1.280$

correlation 0.965 $\quad e$ = distributional residuals, trial weight $w = 0.2$

$n = 99$ $\quad\quad \Sigma|e| = 17.689$

p	x	$\hat{N}(p)$	$\|e\|$	e	$S2^*(p)$
0.007	3.364	3.183	0.181	0.181	-1.752
0.017	3.865	3.613	0.252	0.252	-1.140
0.027	3.932	3.862	0.070	0.070	-1.653
0.037	4.137	4.043	0.094	0.094	-1.419
0.047	4.171	4.187	0.015	-0.015	-1.734
etc.					

Additional model — Pareto correlations 0.930, correlation 0.990, weight = 0.196

$\lambda = -3.0$ $\quad 5.853 = \lambda$

$\beta = 0.5$ $\quad 1.362 = \eta$

$\quad\quad 0.503 = \beta$ $\quad\quad \Sigma|e| = 13.303$

$S2^*(p)$	$\hat{Pa}(p)$	$\hat{Q}(p)$	$\|e\|$
-1.752	-1.996	3.428	0.064
-1.140	-1.991	3.797	0.068
-1.653	-1.986	4.011	0.079
-1.419	-1.981	4.167	0.031
-1.734	-1.976	4.292	0.121
etc.			

Table 8.9. Illustrative layout for forward selection

$$e_{(r)} = x_{(r)} - \hat{Q}_1(p_r)$$

gives the distributional residuals. It is evident that any model components of importance that are not in the current model must make their
effect through the $e_{(r)}$. In the deterministic situation the residuals can
just be plotted and fitted against new components. There is, however,
a problem here for distributional modelling. The deterministic components would lead to residuals, some positive, some negative that
reflected the shape of the missing components. However, a quantile is
inherently an increasing function, so the link between the residuals
and the new component is not a direct one. The way to address this
issue is to observe that we have in the past sought to add quantile
functions in proportions that add to one. Thus if $S2(p)$ is a missing
component, the next stage of the model will look like

$$Q_2(p) = \lambda + \eta[(1 - \omega)S1(p) + \omega S2(p)]$$

If a purely automated approach is taken, then this model is the
added form of stage 3 above and is used in stage 4 of the procedure
and the process is continued. It is likely that one is only looking for
models with a few distributional components. It is therefore useful to
take a more detailed graphical look at this situation. To do this, note
that $Q_2(p)$ can be re-expressed as

$$Q_2(p) = \lambda + \eta S1(p) + \eta\omega[S2(p) - S1(p)]$$

If our model is right, the data is generated by $Q_2(p)$, so approximately $x_{(r)} = Q_2(p_r)$, but we already have $x_{(r)} = \hat{Q}_1(p_r) + e_{(r)}$. The
distributional residual, $e_{(r)}$, is thus the sample equivalent of the population quantity $\eta\omega[S_2(p) - S1(p)]$. Hence, as a rough approximation
since we are not adjusting or estimating parameters but only using
guestimates, a sample quantile function for $S2(p)$ is given by

$$S2^*(p) = \hat{S}1(p_r) + e_{(r)}/\eta\omega,$$

which corresponds to the population component $S_2(p)$. This is calculated in Table 8.9 using a small weight, on the assumption that we
have correctly identified the main component. A look at the plot in
Figure 8.4(a) of the fitted normal quantile function and the data suggests a long-tailed second component. The plot of $S2^*(p_r)$ against p

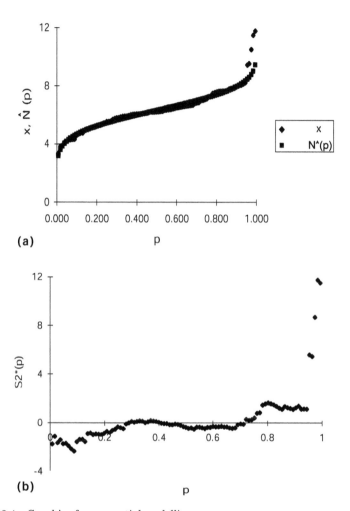

Figure 8.4. Graphics for sequential modelling

supports this. A Pareto, $Pa(p)$, is tried and a plot of a roughly fitted $\hat{S}2(p)$ against $S2^*(p)$ shows a rough linear increase (see Figure 8.4(c)). The corresponding correlation is 0.930, which supports the Pareto as an additional component of the overall distribution. On this basis the full model, $Q(p)$, of weighted normal plus Pareto is fitted and Figure 8.4(d) shows a good fit-observation plot. The least absolutes value shown in Table 8.9 has reduced from 17.7 to 13.3 and the correlation increased from 0.965 to 0.990. It will be seen from this example that there is a forward selection procedure for the distribu-

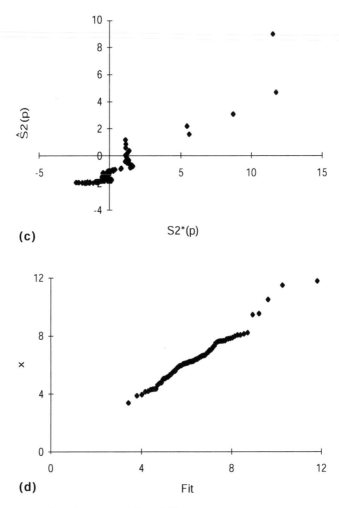

Figure 8.4. Graphics for sequential modelling

tional element of a model that broadly follows the common modelling approach for the deterministic element.

8.6   Problems

1.   Using the population plots of $G(p)$ and $T(p)$ obtained in Chapter 3 problems, suggest suitable models for the two

sets of data for which samples $g(p)$ and $t(p)$ were derived in Chapter 2 problems.

2. Apply the methods of this chapter to Table 2.2 data.

3. Show that the statistic $k(p) = f_p(p)/f_p(0.5)$ is independent of position and scale. Examine its form for some standard distributions. How might a sample measure of this be derived from data? Consider the use of this statistic in identification.

CHAPTER 9

Estimation

9.1 Introduction

As was described in the introduction to modelling in Section 1.10, a major task in statistics is the matching of a model to a set of data. The parameters of the model have to be estimated in such a way as to give a good fit between data and fitted model. The question that has to be asked is, "What do we mean here by good?" There are many answers to this question and hence there are many different approaches to the process of estimation. It is convenient to look at the methods to be discussed in this chapter as two general types. First, there are methods that seek to match specific population properties of the fitted model with the corresponding sample properties of the data. The low order population moments of the fitted model may, for example, be made equal to those of the sample. Second, there are methods based on minimizing some measure of the discrepancy between the fitted model and the data. In Chapter 1, this approach was illustrated with the minimization of the discrepancy measured by the sum of squares or absolute values of the distributional residuals. Most of the common methods of estimation have been developed for use with distributions defined by their density function or cumulative distribution function. We need to show how to fit a model defined by its quantile function. In the process of doing this, we will discover that some methods are particularly suited to distributions defined in terms of $Q(p)$.

9.2 Matching methods

The method of percentiles, illustrated in Chapter 1, equates population and sample percentiles to obtain a number of equations that are solved to provide estimates of the parameters of a distribution. Clearly, the number of equations has to match the number of parameters, k. To generalize, consider a set of descriptors of a population's properties, $G_i(\underline{\theta})$. These will depend on the model, $Q(p;\underline{\theta})$, and its parameters, denoted by $\underline{\theta}$. Corresponding to these population quantities there will be sample quantities, g_i, depending only on the data. The set of equations obtained by matching the population and sample measures will be

$$G_i(\underline{\theta}) = g_i, \; i = 1, \, 2, \, ..., \, k.$$

These can be solved to give the estimates $\underline{\theta} = \hat{\underline{\theta}}$ and the fitted model $\hat{Q}(p)$ given by $Q(p;\hat{\underline{\theta}})$. There are a variety of measures of the shape of a distribution that may be used. The first few moments provide the basis for the method of moments. The use of selected percentiles gives the method of percentiles, sometimes referred to as the method of quantiles, e.g., Bury (1975). In more recent times methods based on L-moments and probability-weighted moments have been used, e.g., Landwehr and Matalas (1979), and Hosking (1990). Table 9.1 lays out the main formulae for these methods, which will now be illustrated through examples. The method of percentiles was also illustrated in Examples 1.24 and 1.25.

Method	$G_i(\theta)$	g_i
Moments	μ'_i or μ_i	$\Sigma x^i/n$ or s^2, m_3, etc.
Percentiles/quantiles	$Q(p_i)$ or M, IQR, QD, etc.	$\tilde{Q}(p_i)$ or m, iqr, qd, etc.
Probability-weighted moments	$\omega_{r,s}$ usually $r = i$, $s = 0$, or $r = 0$, $s = i$	$w_{r,s}$
L-moments	λ_i	l_i

Table 9.1. Matching methods of estimation

Example 9.1: For the Cauchy distribution we have

$$Q(p) = \lambda + \eta \, \tan[\pi(p - 0.5)]$$

If the sample p and q quantiles, $q = 1 - p$, $p < 0.5$, are $x_{(p)}$ and $x_{(q)}$, then the method of percentiles chooses the parameters to equate the population values with the sample values. The two equations $Q(p) = x_{(p)}$ and $Q(1 - p) = x_{(q)}$ are to be solved. Notice first that

$$Q(1 - p) = \lambda - \eta \, \tan[\pi(p - 0.5)],$$

hence by adding it is evident that

$$\hat{\lambda} = (x_{(p)} + x_{(q)})/2.$$

Subtracting the two basic equations gives

$$x_{(q)} - x_{(p)} = -2\hat{\eta} \, \tan[\pi(p - 0.5)],$$

which gives the estimator of η. As a numerical example suppose that from a set of data the two quartiles are 11.6 and 25.4. Putting $p = 0.25$ we have $\hat{\lambda} = 18.5$ and $\hat{\eta} = 6.9$.

Example 9.2: Some of the simplest distributional models are of the form $\lambda + \eta S(p)$, where $S(p)$ contains no parameters. If $S(p)$ is symmetric, then $S(0.5) = 0$ and equating population and sample medians leads to the method of percentiles estimator $\hat{\lambda} = m$. It often happens that results of methods are developed in an ad hoc fashion to provide further estimators. A weakness seen in this estimator is that, for n odd, m is just the central observation. For symmetric distributions the average of any symmetrically placed quantiles will also be λ. On this basis the following form of estimator has been used:

$$\hat{\tau} = \gamma \tilde{Q}(\alpha) + (1 - 2\gamma)m + \gamma \tilde{Q}(1 - \alpha).$$

For $\alpha = \gamma = 0.25$, $\hat{\tau}$ is called the **Trimean**, for $\alpha = 1/3$, $\gamma = 0.3$ it is the **Gastworth** estimator (Gastworth and Cohen (1970).) Parallel to these, percentile estimators of η are given by

$$\hat{\eta} = [\tilde{Q}(1 - \alpha) - \tilde{Q}(\alpha)]/[S(1 - \alpha) - S(\alpha)].$$

If, for example, $S(p)$ is normal, then the estimators of σ corresponding to the Trimean and Gastworth values are

$$\hat{\sigma} = 0.7413 IQR \text{ and } \hat{\sigma} = 1.1608[\tilde{Q}(2/3) - \tilde{Q}(1/3)].$$

See, for example, Srivastava et al. (1992).

Example 9.3: Consider fitting members of the reflection family first introduced in Section 3.6 and for which we have now seen many examples. The form of the distributions is

$$Q(p) = \lambda + (\eta/2)[(1 + \delta)S(p) - (1 - \delta)S(1 - p)].$$

If we take some low value of p ($p < 0.5$), then we can define the population median, M, $IPR(p)$, and p-difference. The subscript S denotes these quantities for the right-tail distribution, $S(p)$, and lower case symbols are the sample values. With this notation we have

$$Q(1 - p) = \lambda + (\eta/2)[(1 + \delta)S(1 - p) - (1 - \delta)S(p)].$$

We could at this stage put sample values on the left-hand sides of the previous three equations and solve for the three unknown parameter values. However, it is clearer if we derive the two common statistics, IPR_Q and PD_Q, from the values of $Q(p)$ and $Q(1 - p)$. We showed in Example 3.11 that

$$M_Q = \lambda + \eta\delta\, S(0.5), \quad IPR_Q = \eta\, IPR_S \text{ and } PD_Q = \eta\delta\, PD_S.$$

The values of IPR_S and PD_S are derived directly from the function $S(p)$. The method of percentiles now replaces the population values of M_Q, IPR_Q, and PD_Q by their sample values to obtain three equations for the three parameters. Thus if we use $p = 1/4$, for the quartile measures we would have

$$m = \hat{\lambda} + \hat{\eta}\hat{\delta}M_S.$$
$$iqr = \hat{\eta}IQR_S.$$
$$qd = \hat{\eta}\hat{\delta}QD_S.$$

From these the following estimators are obtained:

$$\hat{\eta} = (iqr)/IQR_S.$$
$$\hat{\delta} = (qd/iqr)/(QD_S/IQR_S) = g/G_S.$$
$$\hat{\lambda} = m - M_S(qd/QD_S).$$

By way of example, for the skew logistic $S(p)$ is the standard exponential distribution. For this distribution,

$$M_S = \ln 2, \quad IQR_S = \ln 3 \text{ and } QD_S = \ln(4/3).$$

If it is appropriate for the problem, the quartile-based measures can be replaced by those based on some other percentage, perhaps chosen to give a better fit in the tails and to match up with observation points.

Suppose now that we have an $S(p)$ that contains a shape parameter, β, as in, for example, the four-parameter lambda. For this reflected model, the tail index is $T_Q(p) = T_S(p:\beta)$ and this does not involve the other three parameters. Thus some convenient pair $x_{(r)}$ and $x_{(n+1-r)}$ are chosen and p set at p_r. The shape parameter is then chosen so that sample and population values of the corresponding tail indices are equal, i.e.,

$$(x_{(n+1-r)} - x_{(r)})/iqr = T_S(p_r:\hat{\beta}).$$

As this estimate of β is derived independently of the other parameters, it can be then be used in the previous formulae for their calculation, which now depend on β.

In general, the method of percentiles does not produce as good estimates as some of the others to be considered, basically because the estimates do not use all the data. There are, however, some methods to improve the estimates obtained; see, for example, Castillo and Hadi (1994). The method of percentiles is, however, more robust than many others in its ability to cope with erroneous observations that lie outside the rest of the data.

> **Example 9.4:** Illustrating now the method of moments, for the exponential distribution the population mean is in fact the scale parameter η. If we have a sample with mean $\bar{x}$, then the method of moments equates fitted and sample values to give the estimator $\hat{\eta} = \bar{x}$.

> **Example 9.5:** The book on statistical modelling by Shapiro and Gross (1981) gives the formulae for the first four central moments of the generalized lambda distribution, then use the method of moments to estimate the parameters. As there are two non-linear parameters the method is complex, and they provide sets of tables to facilitate the process. The availability of general-purpose numerical procedures avoids the use of such tables and the method becomes straightforward.

There are two basic problems with using the method of moments on distributions like the generalized lambda with three or more parameters. First, the equations are complex and provide many opportunities for error. Second, and more important, the sampling variability of powers of x, like x^3 and x^4, particularly in long-tailed distributions, is very large. Hence estimators based on such quantities are themselves subject to large variability. The methods of L-moments and probability-weighted moments provide alternatives that avoid the problems due to powers of x. It was shown in Chapter 3 that these methods are equivalent. The calculation of probability-weighted moments turns out to be relatively straightforward for the basic models considered in previous chapters. As a consequence, the use of the method of probability-weighted moments is often preferred.

Example 9.6: In Section 4.9 we showed that for the power distribution with known origin at $x = 0$, $Q(p) = \eta p^{\beta}$, the $(r, 0)$ PWM is

$$\varpi_{r,0} = \eta/(r + \beta + 1).$$

Suppose we have a sample of observations on such a power distribution. As there are two parameters, η and β, we need only two sample PWMs, the simplest being $w_{0,0}$ and $w_{1,0}$. Thus we create two equations for the two parameters. This is done by equating sample and population PWMs giving

$$w_{0,0} = \text{fitted } \varpi_{0,0} = \hat{\eta}/(\hat{\beta} + 1),$$

$$w_{1,0} = \text{fitted } \varpi_{1,0} = \hat{\eta}/(\hat{\beta} + 2).$$

The sample PWMs are calculated as in Chapter 3. Solving these equations and simplifying gives

$$\hat{\eta} = w_{0,0}\, w_{1,0}/(w_{0,0} - w_{1,0}),$$

$$\hat{\beta} = (2w_{1,0} - w_{0,0})/(w_{0,0} - w_{1,0}).$$

9.3   Methods based on lack of fit criteria

In Section 1.10 the methods of distributional least squares and distributional least absolutes were briefly introduced. There are a range of methods based on developing some measure of **lack of fit**. The model parameters are then chosen to minimize this criterion. For fitting distributions the natural criteria to use are based on the distributional residuals. We can define these in general terms as the deviations between the ordered observations and some measure of their position derived from the fitted model. These positions we have previously described by the rankit, $\mu_{(r)}$, i.e., the population mean of the r-th ordered observation and the median rankit, M_r, the corresponding population median. Using these as the basis for measures of how well the fitted model describes the data gives two simple criteria:

$$C_{DLS} = \Sigma_r \, (x_{(r)} - \hat{\mu}_{(r)})^2,$$

the **distributional least squares** criterion,

and

$$C_{DLA} = \Sigma_r | \, x_{(r)} - \hat{M}_r | \, ,$$

the **distributional least absolutes** criterion.

Here $\hat{\mu}_{(r)}$ and $\hat{M}_r$ are the rankit and median rankit calculated using the model with estimated parameters. In each case a perfect fit gives zero values of C and a best fit can be obtained by minimizing the criterion by choice of parameters. The minimum magnitudes of the criterion can also be used to compare the best fit to a set of data from one model with that obtained from another.

The method of distributional least squares requires the calculation of the rankits for the given distribution. As we saw in Section 4.2 these are usually not easily calculated, although there is a substantial body of literature giving formulae and tables. The exact rankits are available for such distributions as the exponential, Pareto and power. Approximate rankits are obtained from the formula:

$$E(X_{(r)}) = \mu_{(r)} \approx Q(r/(n + 1))$$

This is a good approximation except in the extreme tails of the distribution $r = 1, 2, 3, \ldots$ and $n, n - 1, n - 2, \ldots$.

As shown in Section 4.5 the use of the next term in the approximation gives the approximate rankit as

$$\mu_{(r)} = Q(p_r) + p_r q_r Q''(pr)/[2(n + 2)],$$

where $p_r = r/(n + 1)$, $q_r = 1 - p_r$, and $Q''(p)$ is the second derivative of $Q(p)$.

Example 9.7: For any distribution of the form $Q(p) = \eta S(p)$, where there is only the one parameter and the rankits of $S(p)$ are $\mu_{(r)}$, then the criterion is

$$\Sigma r[\, x_{(r)} - \eta \mu_{(r)}]^2.$$

Differentiating with respect to η and equating to zero to get the estimator $\hat{\eta}$ gives

$$\Sigma_r[x_{(r)} - \hat{\eta} \mu_{(r)}]\mu_{(r)} = 0$$

solving gives

$$\hat{\eta} = \Sigma_r[x_{(r)}\mu_{(r)}]/\Sigma_r[\mu_{(r)}]^2.$$

Example 9.8: Developing the ideas of this last example consider the fitting of the general model of the form

$$Q(p) = \lambda + \eta S(p;\beta).$$

We can also express the model in terms related to the criteria by expressing the order statistics by

$$x_{(r)} = \lambda + \eta\mu_{(r)} + e_{(r)},$$

where $e_{(r)}$ represents the deviations from the expectation of $X_{(r)}$, $\lambda + \eta\mu_{(r)}$. With the estimated parameters we obtain the distributional residuals, e_r, which are the deviations of the ordered observations from their estimated mean (or median). The criterion now requires that the three parameters are chosen to minimize $\Sigma(e_r)^2$. Using calculus to find the minimum for fixed β gives some results that are standard in the area of regression (see Chapter 12).

$$\hat{\eta} = \Sigma_r[x_{(r)}(\mu_{(r)} - \bar{\mu})]/\Sigma_r[\mu_{(r)} - \bar{\mu}]^2$$
$$\hat{\lambda} = \bar{x} - \hat{\eta}\bar{\mu},$$

where $\bar{\mu}$ here is the average of the rankits and $\bar{x}$ is the average of the data. The minimum value for the criterion is thus

$$C_{min} = \Sigma(x_{(r)} - \bar{x})^2 - \Sigma_r[x_{(r)}(\mu_{(r)} - \bar{\mu})]^2/\Sigma_r[\mu_{(r)} - \bar{\mu}]^2.$$

We clearly want to choose β to minimize this. As the first term depends only on the data, this is equivalent to maximizing

$$T = \Sigma_r[x_{(r)}(\mu_{(r)} - \bar{\mu})]/\Sigma_r[\mu_{(r)} - \bar{\mu}]^2.$$

The optimum β is then used in a re-evaluation of λ and η. The optimum β can also be obtained by directly minimizing $\Sigma(e_r)^2$. Table 9.2 shows a formulation of the steps in the calculation.

Example 9.9: Almost all statistical software and spreadsheets enable linear models with more than two linear parameters to be fitted by a least squares procedure. To illustrate this consider a skew logistic distribution which can be written as

Step	Data in columns	Derived quantities	Notes
1	set β at initial value		Arbitrary value, can be explored later
2	ordered data $x_{(r)}$	$\Sigma x_{(r)} = S_x$	S = Sum, SS = Sum of Squares, SP = Sum of Products, D = Deviations
3	deviations $x'_{(r)}$		$= x_{(r)} - \Sigma x_{(r)}/n$
4	rankits $\mu_{(r)}$	S_μ	$= \Sigma \mu_{(r)}$
5	deviations $\mu'_{(r)}$	SSD_μ	$= \Sigma(\mu_{(r)} - S_\mu/n)^2$
6	$x'_{(r)} \mu'_{(r)}$	$SPD_{x\mu}$	
7		$\hat{\eta} = $ $SPD_{x\mu}/SSD_\mu$	
8		$n\hat{\lambda} = S_x - \hat{\eta} S_\mu$	
9	fitted $\hat{x} = \hat{\lambda} + \hat{\eta}\mu_{(r)}$		
10	dist res $= e_r$ $= x_{(r)} - \hat{x}_{(r)}$	SS_e	
11		Adjust β to minimize SS_e	Software should automatically alter $\hat{\lambda}$ and $\hat{\eta}$

Table 9.2. Layout of DLS calculation for $Q(p) = \lambda + \eta S(p;\beta)$

$$x_{(r)} = \lambda + \eta\ln\{p_r/(1 - p_r)\} - \eta\delta\ln\{p_r(1 - p_r)\} + e_{(r)}$$

The product $\eta\delta$ can be treated as a new single parameter θ. The new three linear parameters can be estimated using software to carry out a "regression" of $x_{(r)}$ on $c_{(r)}$ and $d_{(r)}$ where $c_{(r)} = \ln\{p_r/(1 - p_r)\}$ and $d_{(r)} = -\ln\{p_r (1 - p_r)\}$; thus fitting by standard least squares the model

$$x_{(r)} = \lambda + \eta\, c_{(r)} + \theta d_{(r)} + e_{(r)}.$$

Turning now to distributional least absolutes criterion, C_{DLA}, we should explain why we have used the median rankit in the absolute value criterion. Part of the reason for this usage is a general result that for any set of observations, x_j, the quantity

$$\Sigma|\, x_j - a\,|$$

is minimized by setting a at the median value of the x_j. Notice that here the median may be the single central value or any number between the pair of central values, so the median value is not necessarily unique.

The median rankit is the natural measure of middleness in the least absolutes criterion.

Example 9.10: Suppose $S(p)$ is a distribution with range $(-\infty, \infty)$ and the model is

$$Q(p) = \lambda + S(p).$$

The DLA criterion is thus

$$\Sigma|x_{(i)} - \lambda - S(p_i^*)|.$$

By the above result we obtain

$$\hat{\lambda} = \text{Median}(x_{(i)} - S(p_i^*)), \ i = 1, \ldots, n.$$

Notice that the median rankits are given by $Q(p_r^*)$ as an exact universal result, unlike the corresponding problem of finding the exact or approximate rankits required for the DLS criterion.

Example 9.11: Table 9.3 illustrates the fitting of a set of data using distributional least absolutes. Initial values of the parameters are allocated and the median rankits calculated. The distributional residuals, e_r, are then found and the least absolutes criterion evaluated. The soft-

Generalized	Lambda	$\hat{\lambda} = 60.84$		
$n = 100$		$\hat{\eta} = 96.05$		
		$\hat{\alpha} = 0.21$		
		$\hat{\beta} = 0.02$		
		$\Sigma|e_r| = 114.43$		
r	x	p_r^*	$M_r = \hat{Q}(p_r^*)$	$|e_r|$
1	7.49	0.007	−0.64	8.13
2	8.11	0.017	6.27	1.84
3	13.43	0.027	10.45	2.98
4	13.55	0.037	13.55	0.01
5	14.54	0.047	16.04	1.50
		etc.		

Table 9.3. Fitting by the method of least absolutes

ware then searches to find the set of four parameters to minimize the criterion. Table 9.3 shows the set obtained at this final stage. The general algorithm for both DLS and DLA is given in the next section.

Fitting by distributional least squares and distributional least absolutes has been discussed in relation to raw data. If there is only data in frequency table form with class intervals (x_{i-1}, x_i) and cumulative relative frequency p_i at x_i (i.e., the empirical CDF is $\tilde{F}(x_i) = p_i$), then the error of fit term will look like

$$e_i = x_i - Q(p_i;\theta)$$

The parameters, θ, can be chosen to minimize the least squares or least absolutes criteria. The nature of a frequency table is that these criteria give equal weight to each class interval (x_{i-1}, x_i) irrespective of the frequency, f_i, in the intervals. For fitting with emphasis on a good fit in the tails this is reasonable. If all observations should make an equal contribution, then the criteria are weighted by the frequency, e.g., Σe_i^2 becomes $\Sigma f_i\, e_i^2$. For this criterion we cannot minimize using explicit algebraic expressions, as for simple least squares. We have to use numerical optimization.

Having introduced these methods of estimation we need to consider their properties. Simple least squares is the classical method of estimation. Its use does, however, depend on a number of assumptions: If the criteria takes the form $\Sigma(x_i - \mu_i)^2 = \Sigma(e_i)^2$, then it is assumed that the residuals e_i (a) are independent and (b) have constant variance. A further assumption implied in much of the use of least squares is that the residuals are (c) from a distribution that is symmetrical. It is clear that when the x are the order statistics none of these assumptions are true. The method of least squares can be generalized to deal with (a) and (b). The theory was first developed for the order statistics by Lloyd (1952) (see also Arnold, Balakrishnan and Nagaraja (1993)). In essence, the approach involves constructing a matrix of all the variances and covariances for the order statistics, which is beyond the scope of this text. Balakrishnan and Sultan (1998) give appropriate formulae for such quantities for a wide range of distributions. For a data set of any size the calculation of all the n^2 variances and covariances becomes difficult even if these quantities are explicitly known, and this is only the case for some distributions. The task becomes even more difficult where recurrence relations or numerical approximations are required to obtain these quantities. If we ignore the correlation structure and concentrate on the variability, we can make some progress. The standard approach to this problem is based on weighting the terms in the criteria, which becomes for the least squares criterion:

$$C_{DWLS} = \Sigma_r\, \omega_r (x_{(r)} - \mu_{(r)})^2,$$

distributional weighted least squares.

The usual form for the DWLS criteria uses the variances of the order statistics to reduce the effect of their errors. Thus $\omega_r = 1/V(X_{(r)})$. For example, the formula for an approximate variance for the normal distribution is given in the discussion in Section 5.6. For DWLA the equivalent criterion is

$$C_{DWLA} = \Sigma_r\, \omega_r |x_{(r)} - M_r|,$$

distributional weighted least absolutes,

where ω_r is the reciprocal of some suitable quantile measure of spread, such as a suitable interquantile range. The inter-decile range, *IDR*, is a sufficiently wide but not too extreme range to use. Unlike the variance, this can be obtained exactly from the quantile function. Thus we have

$$M_r = Q[p_r^*(0.5)]$$
$$IDR_r = Q[p_r^*(0.9)] - Q[p_r^*(0.1)]$$
$$= Q[\mathrm{BETAINV}(0.9, r, n + 1 - r)]$$
$$- Q[\mathrm{BETAINV}(0.1, r, n + 1 - r)].$$

The weighting used is then, for a given fitting, $\omega_r = 1/I\hat{D}R$. Table 9.4 shows a column layout of part of such a calculation for fitting a generalized lambda distribution. The modification to the standard approach is a simple multiplication by weighting before summing to get the criterion value. Clearly, the weighting depends on the parameters, but this is allowed for automatically when the parameters are chosen to minimize the criterion.

There is some evidence that for ordinary least squares ignoring the issue of the variances and correlations does not dramatically worsen the results (see, for example, Ali and Chan (1964)). It has to be noted that the outcome of the least squares criterion does indeed minimize the criterion. It is consideration of the theoretical properties of the estimates obtained that leads to the use of weightings and adjustment for covariance. The modeller has to balance the issues as to whether

Generalized Lambda

$n = 100$

$$\hat{\lambda} = 58.70 \qquad e_r = x_r - \hat{Q}(p_r^*(0.5))$$

$$\hat{\eta} = 329.06 \qquad 1/w_r = \hat{Q}(p_r^*(0.9)) - \hat{Q}(p_r^*(0.1))$$

$$\hat{\alpha} = 0.048$$

$$\hat{\beta} = 0.009$$

$$\Sigma w_r |e_r| = 15.88$$

r	x_r	$p_r^*(0.1)$	$p_r^*(0.5)$	$p_r^*(0.9)$	$\hat{Q}(p_r^*(0.1))$	$\hat{Q}(p_r^*(0.5))$	$\hat{Q}(p_r^*(0.9))$	$w_r|e_r|$
1	7.49	0.001	0.007	0.023	−34.04	−11.56	3.83	0.50
2	8.11	0.005	0.017	0.038	−14.78	−0.24	10.87	0.33
3	13.43	0.011	0.027	0.052	−5.58	5.94	15.17	0.36
4	13.55	0.018	0.037	0.066	0.04	10.23	18.33	0.19
5	14.54	0.025	0.047	0.078	4.82	13.54	20.86	0.06

etc.

Table 9.4. Fitting by the method of weighted least absolutes

the objective is to obtain estimates of the parameters that have statistically good properties, assuming all is well with the choice of model
and with the data, or is it to minimize a practical meaningful criterion
of fit.

The lack of symmetry in the distributions of the outer order
statistics also raises issues. The skewness and high variability in
these tails mean that squaring the e_i leads to estimates that are very
sensitive to this variability and to any outliers. This problem is
clearly reduced by using DLA rather than DLS and also by introducing a weighting, ω_r. However, one could argue that the equal weighting that ordinary DLS and DLA place on the data in the tails may
not always be unreasonable, since we often wish to obtain the best
fit in the tails.

In practical modelling studies it should be the application that
determines the criterion, guided by, but not determined by, statistical
theory. In many applications the distributional least absolutes criterion turns out to be a natural measure of fit. This criterion and the
equivalent least absolutes criterion for fitting deterministic models
have been much discussed. A significant problem is that the criteria
do not lead to unique solutions. There will be a small range of values
of the parameters for which the criterion does not change. To see this,
consider applying the criteria when the model is $Q(p) = \lambda + S(p)$, where
$S(p)$ is symmetrical about zero. A natural estimate of λ is the median.
If, however, there are an even number of observations the criterion is
constant for values of λ lying between the two middle observations.
Although this is clearly a fact, under the appropriate circumstances
in practice with a reasonably sized sample and several parameters,
the range of variation in the estimates is practically insignificant. It
is also open to simple exploration, to see whether for any situation it
is an issue.

In favour of the DLA criterion it may noted that when the distributions are not symmetrical a criterion based on median rankits is
intuitively understandable. Further, as we have seen, DLA, compared
with DLS, is relatively insensitive to outliers and problems of high
variability. Teaching texts on statistics tend to put great emphasis on
least squares, often to the exclusion of all else. Just to emphasize that
there is a choice in this matter and that other criteria should be
considered, we will use distributional least absolutes as the criterion
for fitting in the remainder of this text.

Although DLA is more robust than DLS it is still influenced by
outliers. One way of constructing a highly robust approach is to note
first that both criteria are equivalent to minimizing the mean of the

squares or absolute values of the distributional residuals. If we replace the sample mean by the sample median we have methods called the distributional least median of squares and of absolutes. The least median of squares method was introduced by Rousseeuw (1984) and shown by him to be a highly robust method. From the point of view of obtaining numerical optima against some criterion of fit, this criterion is as easy to handle as the least means criteria.

It is shown in many standard texts that the method of least squares has the effect of maximizing the square of the correlation between the actual and fitted observations, which is called the **multiple correlation coefficient**. This coefficient will lie just below its possible maximum of one for any sensible model. The fact that it has a known range makes it useful as a supplementary measure of the quality of fitted models however they are derived.

The multiple correlation coefficient and the DLS and DLA criteria give overall measures of the fit of a distribution. When we ask about the quality of the estimates of the individual parameters we usually have little simple information. If a linear model with variances and covariances is used for distributional least squares, then formulae for the standard errors of the estimates of the linear parameters are available. For non-linear parameters and for the method of distributional least absolutes there are rarely simple formulae available. The approach increasingly adopted to such issues is to repeatedly resample, with replacement, from the original data set and study the variability of the estimators obtained.

9.4   The method of maximum likelihood

The basic idea of **likelihood** is that for an observation x_i the probability of occurrence in a small interval dx_i at x_i is $f(x_i;\theta)\ dx_i$, where $f(x;\theta)$ is the PDF with parameter(s) θ. For a set n independent observations this probability becomes $\Pi[f(x_i;\theta)\ dx_i]$. Hence a general measure of the chances of seeing the particular observed set of data, seen as a function of the parameter, is given by the likelihood defined as

$$\text{Likelihood}(\theta) = \Pi f(x_i;\theta).$$

Intuition suggests that if a value of θ is used that is close to the true value then this quantity will be relatively large; if we have a poor choice, it will be small.

In general, the likelihood of a parameter θ, given a set of data x_i from a distribution, $f(x{:}\theta)$, is defined by

$$L(\theta) = f(x_1{:}\theta)\, f(x_2{:}\theta)\, \dots\, f(x_n{:}\theta).$$

We can define the equivalent likelihood quantile function

$$L(\theta) = f_p(p_{(1)}{:}\theta)\, f_p(p_{(2)}{:}\theta)\, \dots\, f_p(p_{(n)}{:}\theta)$$

where $p_{(r)}$ is given by $x_{(r)} = Q(p_{(r)}{:}\theta)$, i.e., it is the actual p value that generates the observed x for the given θ. It is easier to work with the **log-likelihood** which we will denote by $\ell(\theta)$.

$$\ell(\theta) = \ln[L(\theta)] = \Sigma\ln f_p(p_{(r)}{:}\theta).$$

Expressing this in terms of the quantile density gives

$$\ell(\theta) = -\Sigma\ln q(p_{(r)}{;}\theta).$$

The use of logarithms enables the terms to be separately studied and also provides for $\ell(\theta)$, for a given θ, to be evaluated as a column total, for example, on a spreadsheet. The log-likelihood acts not as a lack of fit measure, but as a goodness-of-fit criterion, $C_{ML} = \ell(\theta)$, that has to be maximized.

The method of maximum likelihood chooses the parameter(s) to maximize the likelihood or, equivalently, the log-likelihood. This is probably the most commonly used method of estimation. The reason for its popularity is that it leads to estimators that usually have particularly good and useful general properties. In summary, the main properties are

(a) The maximum likelihood estimators, $\hat{\theta}$, for large samples are approximately normally distributed, unbiased (i.e., $E(\hat{\theta}) = \theta$), and with a variance that is at the minimum achievable value.

(b) If the unbiasedness and minimum variance are theoretically achievable in small samples, then they will be given by maximum likelihood.

(c) The variance properties of the estimators can be directly derived.

(d) The estimators become closer in a probabilistic sense to the true values as the number of observations increases.

The technical term is that they are **consistent** estimators. These properties are usually valid for distributions that are straightforward functions of the parameters.

The calculation of $\ell(\theta)$ requires the derivation of $p_{(r)}$ from $x_{(r)} = Q(p_{(r)}:\theta)$. This can be carried out directly from $F(x_{(r)})$ if $F(.)$ is explicit. If not, the iterative method of Section 4.5(c) has to be used. It is seen that in the quantile-based likelihood the parameter(s) are implicitly contained in terms involving $p_{(r)}$. This adds a fresh level of complexity to the situation. Thus the methods for quantile functions discussed here are only needed for distributions that have quantile functions but not explicit probability density functions.

We have discussed three criteria of fit: C_{DLS}, C_{DLA}, and C_{ML}. The model parameters, $\underline{\theta}$, are chosen to either minimize or maximize the criterion $C(\underline{\theta})$. In most standard statistics texts these criteria, where discussed, are optimized analytically, giving explicit answers. With quantile-based models this is rarely possible and a universal algorithmic approach is needed. We will see later when we turn to validation that this approach has benefits. The algorithm can be set up via spreadsheets or as a procedure in a statistical package. Table 9.5 shows the structure of the algorithm. The layout also calculates a number of quantities that are of later use.

The approach of this table emphasizes the need to decide in any problem what model to use and what criteria of fit to use. These are quite distinct issues. In the past the criterion has often been chosen on the basis of obtaining simple formulae for solution, or obtaining good statistical properties for the estimates. The second of these is clearly sensible, but it should not override the purposes of the model. If we make use of general-purpose minimization/maximization routines in spreadsheets or mathematical and statistical software, then we have a general tool without need for specific formulae to obtain solutions. As with all such tools care needs to be taken to check the validity of the solutions obtained. Section 9.6 looks at some methods that assist in this.

Before leaving our discussion of the main methods of estimation, it is worth noting that the generalized lambda distribution is the only distribution defined explicitly by a quantile function that has a literature on its estimation. This covers the method of moments, Dudewicz et al. (1974), Ramberg et al. (1979) and Karian et al. (1996); the method of least squares, Ozturk and Dale (1985); the method of percentiles, Karian and Dudewicz (1999a and b); and the method of probability-weighted moments, Greenwood et al. (1979). The lack of use of maximum likelihood is surprising as it is perfectly straightforward if one uses the

1.	Define distribution	$Q(p;\underline{\theta})$
2.	Derive quantile density	$q(p;\underline{\theta})$
3.	Set initial or revised parameter	$\hat{\underline{\theta}}$ ($= \underline{\theta}_0$ on first iteration)
	values (Section 9.6)	
4.	DLA/ML, set median-p_r	$p_r^* = \mathrm{BETAINV}(0.5, r, n+1-r)$,
	or initial $p_{(r)}$	$\mu_{(r)}$ from formulae or tables,
		$p_{(r),0} = r/(n+1)$
5.	ML, derive $p_{(r)}$ (where $x_{(r)} =$	iterate for each r and current $\hat{\underline{\theta}}$
	$Q(p_{(r)};\hat{\underline{\theta}})$); DLA/S, derive M_r	$p_{(r)\mathrm{new}} =$
	or $\mu_{(r)}$	$p_{(r)} + (x_{(r)} - Q(p_{(r)};\hat{\underline{\theta}}))/q(p_{(r)};\hat{\underline{\theta}})$
		$M_r = Q(p_r^*;\hat{\underline{\theta}})$
		$\mu_{(r)}$ from formulae or tables
6.	Calculate criterion	$C(\mathrm{data}\ x_{(r)}|p_{(r)};\hat{\underline{\theta}})$
7.	Search for $\hat{\underline{\theta}}$ to optimize 6	$\underline{\theta} = \hat{\underline{\theta}}$, optimum $C = C(\hat{\underline{\theta}})$.
	(involving iterating steps 3 to 6)	
8.	Fitted distribution	$\hat{Q}(p;\hat{\underline{\theta}})$
9.	Fitted quantile density	$\hat{q}(p;\hat{\underline{\theta}})$

Table 9.5. Algorithm for least/maximum methods

general-purpose maximization software available rather than look for specific formulae for estimators. Table 9.6 illustrates the layout for maximum likelihood calculations in a column format. This gives a section from a fit of the generalized lambda for a set of 100 observations. Columns 2, 3 and 4 give the data, the median-p and the median rankits, calculated using the initial parameter estimates. Columns (4, 5, 6), (7, 8, 9), ..., (13, 14, 15) give the iterative derivation of the column $p_{(r)}$. It will be noted that, as required, column 16, $\hat{Q}(p_{(r)})$, is identical to the data of column 2. The $p_{(r)}$ are used to get the individual likelihood terms $-\ln(q(p_{(r)}))$ whose column is summed to give the final log-likelihood. The parameters are then chosen to maximize this.

9.5   Discounted estimation

A common feature in fitting distributions for applications is that it is only one end of the distribution that is of practical importance. For example, we may fit a distribution to a population of individuals and want to use the fitted model to select the top 1% of future samples. Thus the value of $\hat{Q}(0.99)$ would be the prime statistic of interest. In such situations it is reasonable to seek a better fit for the distribution

Generalized lambda			$\hat{\lambda}$ = 50.00			
n = 100			$\hat{\eta}$ = 67.38			
			$\hat{\alpha}$ = 0.31			
column			$\hat{\beta}$ = 0.04			
1	2	3	4	5	6	7
r	x_r	p_r^*	$\hat{Q}(p_r^*)$	$\hat{q}(p_r^*)$	$p_{r,1}^*$	$\hat{Q}(p_{r,1}^*)$
1	7.49	0.007	6.96	651.8	0.0077	7.47
2	8.11	0.017	11.54	354.6	0.0071	7.05
3	13.43	0.027	14.52	257.6	0.0224	13.36
4	13.55	0.037	16.81	207.4	0.0209	12.91
5	14.54	0.047	18.70	176.0	0.0229	13.50

etc.

log likelihood = $-\Sigma \ln \hat{q}(p_{(r)})$ = -416.1

7		15	16	17	18
$\hat{Q}(p_r^*,1)$	iteration	$p_{(r)}$	$\hat{Q}(p_{(r)})$	$\hat{q}(p_{(r)})$	$-\ln \hat{q}(p_{(r)})$
7.47		0.008	7.49	601.5	−6.40
7.05		0.009	8.11	550.0	−6.31
13.36		0.023	13.43	287.9	−5.66
12.91		0.023	13.55	284.2	−5.65
13.50		0.027	14.54	256.9	−5.55

etc.

Table 9.6. Layout for fitting by maximum likelihood

at the end of interest rather than a good overall fit, which has been the objective in previous methods. The natural approach to this problem is to weight the various criteria we have been using to emphasize the end of interest. As an illustration, consider the use of **exponential discounting** to emphasize the right-hand data when fitting with least absolutes. The criterion becomes

$$\Sigma \, a^r |\, x_{(n-r)} - M_{n-r}|$$

where the **discounting factor**, a, is in the range $0 \le a \le 1$. The smaller a the less emphasis is given to data away from the right-hand tail. However, a cannot be too small as smaller a effectively reduces the amount of data used in the estimation. Values of $a > 0.95$ are normally required.

Example 9.12: A five-parameter lambda distribution was fitted to a set of nearly 500 observations using the method of distributional least absolutes. The process was repeated using exponential discounting with unit weight on the largest observation. Thus the emphasis was on the right-hand tail, since the fitted model was to be used to put a cutoff point in that tail for the selecting out of high values. With such a large data set, a high value of a was used ($a = 0.995$). This gave a weight of about 0.1 on the smallest observations. Figure 9.1 shows the residual plots for the two fitted models. It will be seen that although there is no dramatic difference, the residuals on the right for the discounted fitting have been reduced at the expense of those on the left.

An obvious problem with this approach is in the choice of a. If the method is to be used repeatedly, then a value can be chosen that would

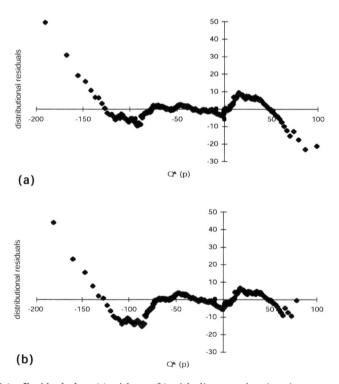

Figure 9.1. Residual plots (a) without; (b) with discounted estimation

have lead to the optimum value of some criteria of success in the past applications. For a "one-off" application a subjective choice has to be made to give reasonable weighting across the data. Such subjective decisions are sometimes frowned upon; however, using the standard method, in effect setting $a = 1$, and ignoring an important aspect of the application, are no more objective than facing the fact of the relative practical importance of the two tails.

9.6   Intervals and regions

The methods discussed in this chapter have all been concerned with obtaining estimated numerical values for parameters. In this section we consider methods for obtaining intervals or regions which in some sense may contain the true value. We start by considering a range of examples to illustrate some basic methodologies.

Example 9.13: Consider the single parameter distribution of form $Q(p) = \eta R(p)$. Let α be a small probability and x a single observation or statistic which has distribution $Q(p)$. By the definition of quantiles

$$Pr[Q(\alpha) \le x \le Q(1 - \alpha)] = 1 - 2\alpha.$$

Rearranging within the probability statement we can write

$$Pr[x/R(1 - \alpha) \le \eta \le x/R(\alpha)] = 1 - 2\alpha.$$

The interpretation of this probability statement is that the random interval $[x/R(1 - \alpha), x/R(\alpha)]$ will contain the true parameter value for $100(1 - 2\alpha)\%$ of observations x. Intervals of this nature are called $100(1 - 2\alpha)\%$ **confidence intervals**. By way of illustration let $R(p) = -\eta \ln(1 - p)$ for the exponential distribution. Suppose the value of $\alpha = 0.025$ and the observed x is 2.2, then the form of the interval comes from $Pr[0.271\, x \le \eta \le 39.50\, x] = 0.95$. The actual 95% confidence interval is $(0.6, 86.9)$. Note the inevitable width of an interval where we ask for a high probability on the basis of only one observation.

Example 9.14: Suppose that some statistic, t, derived from n observations, has a normal distribution with mean μ and known standard deviation σ. We can again write a probability statement with quantile form:

$$Pr[\mu + \sigma N(\alpha) \le t \le \mu + \sigma N(1 - \alpha)] = 1 - 2\alpha.$$

Using the symmetry of $N(p)$ gives

$$Pr[\mu - \sigma N(1 - \alpha) \leq t \leq \mu + \sigma N(1 - \alpha)] = 1 - 2\alpha,$$

which on reorganizing is

$$Pr[t - \sigma N(1 - \alpha) \leq \mu \leq t + \sigma N(1 - \alpha)] = 1 - 2\alpha.$$

If, for example, we note that $N(0.975) = 1.96$, then the 95% confidence interval is

$$(t - 1.96\sigma, \, t + 1.96\sigma).$$

Example 9.15: In Section 5.7 we noted that order statistics are approximately normal for large n. In particular, if $p_r = r/(n + 1)$, then

$$X_{(r)} \text{ is } N[Q(p_r), \, p_r(1 - p_r)q^2(p_r)/n].$$

If $p_r = 0.5$, then $X_{(r)}$ is the sample median m. If the distribution is the logistic with known η, $Q(p) = \lambda + \eta \ln[p/(1 - p)]$, then $q(p) = \eta/[p(1 - p)]$ and $q(0.5)$ is 4η. The median M is λ and hence the sample median is approximately distributed as $N(\lambda, \, 4\eta^2/n)$. Following the same line of argument as in the previous example, we arrive at the 95% confidence interval for λ as

$$[m - 1.96 * 2\eta/\sqrt{n}, \, m + 1.96 * 2\eta/\sqrt{n}].$$

As a specific example suppose $n = 64$, $m = 24$ and $\eta = 2$, then the 95% interval is approximately $(23, 25)$.

Example 9.16: Given the log-likelihood $\ell(\theta)$, the following useful quantities can be defined:

The Score $Sc(\theta) = \partial\ell/\partial\theta$, noting that $Sc(\hat{\theta}) = 0$.

The Information Function $I(\theta) = -\partial^2\ell/\partial\theta^2$.

It can be shown that the variance of the maximum likelihood estimator $\hat{\theta}$ is approximately $1/I(\theta)$. It can also be shown that maximum likelihood estimates are approximately normally distributed about the true value. These facts can be used to obtain approximate confidence intervals for parameters from their maximum likelihood estimates. For example, the 95% confidence interval will be approximately $\hat{\theta} \pm 1.96/\sqrt{I(\hat{\theta})}$.

The use of approximations, as in the previous examples, can be avoided by taking a somewhat different approach. We define two functions: The **likelihood ratio** $R(\theta) = L(\theta)/L(\hat{\theta})$ and the **log-likelihood ratio** $r(\theta) = \ln(R(\theta))$. It is evident that $0 \leq R(\theta) \leq 1$, and that $R(\hat{\theta}) = 1$. A **100p% likelihood interval** or **region** is defined as the region where $R(\theta) \geq p$, or equivalently $r(\theta) \geq \ln(p)$. Given the layout of the ML calculations, such as that of Table 9.6, the estimates can be varied from their maximum likelihood values and the effect on the log-likelihood of so doing can be readily explored.

If the log-likelihood is expanded by a Taylor series about a single parameter $\hat{\theta}$ and we use the functions $Sc(\theta)$ and $I(\theta)$, we obtain

$$\ell(\theta) = \ell(\hat{\theta}) + (\theta - \hat{\theta})\, Sc(\hat{\theta}) + \{(\theta - \hat{\theta})^2/2\}\, I(\hat{\theta}) + \dots.$$

But $Sc(\hat{\theta}) = 0$ for the maximum likelihood estimator and $r(\theta) = \ell(\theta) - \ell(\hat{\theta})$, so as an approximation

$$r(\theta) = -\{(\theta - \hat{\theta})^2/2\}\, I(\hat{\theta}).$$

Thus the interval $r(\theta) \geq \ln(p)$ gives the 100p% interval for θ as

$$\hat{\theta} \pm \sqrt{[-2\ln(p)/I(\hat{\theta})]}.$$

For $p = 0.1$, for example, we would regard the values of θ within a 100p% interval as likely. For $p = 0.147$ the region is approximately the 95% confidence interval. Notice, however, that the likelihood regions in general will not be symmetrically placed about the estimate $\hat{\theta}$. If initial estimates of the end points p_1, p_2 are obtained from the last expression, the equation $r(\theta) - \ln(p) = 0$ can be solved using the approximation methods of Section 4.5. The process of obtaining the 100p% likelihood interval is seen as an extension of the original maximum likelihood estimation and the algorithm for finding the end points is given in Table 9.7.

The distributions considered in past chapters have almost always had at least a position and a scale parameter. In practice therefore we are looking at multiparameter situations. The numerical-based approach that has been discussed for finding maximum likelihood estimates lends itself to a further use. The setup of all the least A and maximum B methods has been the same: first, define starting values for the parameters, $\underline{\theta}_0$, and calculate the criterion, $C(\underline{\theta}_0)$ from

Formulae for derivatives of $\ell(\theta)$	$Sc(\theta)$, $I(\theta)$
Initial values for the end points (θ_1, θ_2)	$\hat{\theta} \pm \sqrt{[-2\ln(p)/I(\hat{\theta})]}$
Iterative evaluation for the two	$\theta_{new} = \theta - [r(\theta) - \ln(p)]/Sc(\theta)$
end points	$\theta = (\theta_1, \theta_2)$
$100p\%$ likelihood interval	(θ_1, θ_2)

Table 9.7. Obtaining $100p\%$ likelihood intervals

these; then search for the parameter values, $\hat{\theta}$, that lead to the minimum or maximum, $C(\hat{\theta})$. Once this is done we can define a lattice of points in the parameter space around the optimum values and evaluate the criterion function at each of these. A criterion ratio will be given by $R^*(\underline{\theta}) = C(\underline{\theta})/C(\hat{\underline{\theta}})$, for a maximum criterion or its reciprocal for a minimum criterion. As with the likelihood ratio, this general ratio will satisfy $0 \leq R^*(\underline{\theta}) \leq 1$, $R^*(\hat{\underline{\theta}}) = 1$. The value of $R^*(\underline{\theta})$ can be directly derived for each point on the lattice using the column layout used for the optimization. The region where $R^*(\underline{\theta}) \geq p$ can then be identified at whatever level of accuracy is desired. Table 9.8 shows the values of the ratio, as a percentage, for the sum of absolutes criterion for a lattice of values of two parameters, η and β, in a model. The 100% value corresponds to the method of least absolutes estimates of the two parameters. Figure 9.2 shows the three-dimensional plot and a contour plot for the actual magnitudes of the DLA criterion. Such exploratory tools give a good feel for the sensitivity of the optimum estimators.

					β				
η	0.33	0.35	0.37	0.39	0.41	0.43	0.45	0.47	0.49
370	31	34	38	44	50	57	59	58	55
380	32	37	43	51	59	67	68	66	59
390	34	40	48	58	70	79	79	73	62
400	36	42	52	65	83	91	87	75	62
410	38	45	57	72	93	97	89	75	59
420	40	49	62	80	100	97	86	70	52
430	42	52	67	89	98	92	78	60	45
440	44	56	71	90	89	80	66	50	39
450	46	58	71	83	77	67	54	43	34
460	48	58	68	72	66	56	45	37	30
470	48	56	63	62	56	47	39	32	29

Table 9.8. Example of % criterion ratio for sum of absolutes criterion

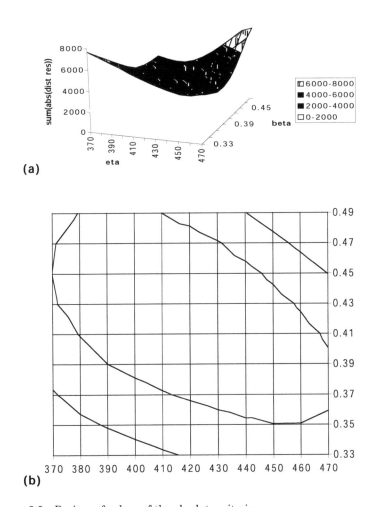

(a)

(b)

Figure 9.2. Regions of values of the absolute criterion

9.7   Initial estimates

Most of the methods of estimation involve the use of numerical opti-
misation routines. These require the setting up of tabular layouts
based on initial estimates. We therefore need to consider briefly the
derivation of these initial values. A number of methods can be used:

(a) The method of percentiles often provides a simple and direct method of getting estimates.

(b) The fitting of simpler but related distributions has proved useful. Suppose, for example, the generalized lambda is reparameterized to

$$Q(p) = \lambda + \{\eta/(\alpha + \beta)\}[p^\alpha - (1 - p)^\beta],$$

and we are confident the shape parameters have the same sign. This distribution can be compared to the symmetric lambda distribution, which, in classical form, is

$$Q(p) = \lambda + (\eta/2\alpha)[p^\alpha - (1 - p)^\alpha].$$

The method of percentiles can give simple estimators of λ, η, and α. We then make $\beta = \alpha$ and fit the four-parameter model from these starting values.

(c) A further approach makes use of the fact that most of the models we have considered have several linear parameters and only one or two non-linear parameters. The non-linear parameters are set at guessed values and simple distributional least squares estimates of the linear parameters derived from the resultant linear model. From this fitted model the criteria of interest can be calculated. The non-linear parameters can then be adjusted by trial and error to give a rough optimum for the criterion. The structures of spreadsheets lend themselves to this type of preliminary exploration.

9.8   Problems

1. The data of Table 2.2 are negatively skewed but positive. The distribution $\lambda - \eta(\ln p)^\beta$ is suggested.
    (a) Justify this form for the distribution.
    (b) Fit the model to the data using the method of percentiles.
    It is suggested that the values 1450 and 1960 are outliers. Fit the normal and power-Pareto distributions using the method of distributional least absolutes to the main 48 observations and consider the situation of the "outliers."

Fit the whole 50 observations using (i) a four-parameter lambda and (ii) a generalized lambda, again using the method of distributional least absolutes.
Review the outcomes of this exercise.

2. Obtain explicit formulae for the estimates of the parameters of the following models using the method of probability weighted moments:
   (a) The logistic, $Q(p) = \lambda + \eta \ln[p/(1-p)]$ in terms of $w_{0,s}$.
   (b) The two-parameter Weibull, $Q(p) = \eta[-\ln(1-p)]^\beta$.

3. A skew four-parameter lambda distribution is fitted to a set of data for which

$$m = 20, \; iqr = 3, \; ipr(0.05) = 6, \text{ and } qd = 1.$$

Estimate the parameters using the method of percentiles.

4. A distribution, $Q(p) = \lambda + \eta S(p)$, has $S(p)$ symmetric about zero. Show that the statistic

$$t = \Sigma_1^n S(r/(n+1))x_{(r)}$$

provides an estimate of η with $E(t) = \eta$, approximately for large n.

5. Show that for distributions of the form

$$Q(p) = \lambda + (\eta/2)[(1+\delta)S(p;\beta) - (1-\delta)S(1-p;\beta)],$$

the right-tail index $\tau = [Q(0.9) - Q(0.5)]/[Q(0.75) - Q(0.5)]$ depends only on β and could thus provide an estimate of β by equating sample and population values. Use this result with the method of percentiles to fit a skew lambda to the data of Table 2.2.

6. Fit the various members of the lambda family to the data of Table 2.2 using the method of maximum likelihood. Compare the maximum likelihoods obtained. For the four-parameter lambda, examine the likelihood region for δ and α, assuming the estimates for λ and η are correct.

7. A distribution is modelled by

$$Q(p) = \lambda + \eta[(2p - 1)/\{p(1 - p)\}]$$

Give formulae for estimating the position and scale parameters λ and η using the method of percentiles and the
method of probability-weighted moments.

8. A distribution is modelled by

$$Q(p) = \lambda + \eta[(1 + \delta)/(1 - p) - (1 - \delta)/p].$$

Calculate M, IQR, T(p), D(p) and G(p). Suggest a simple
way of estimating δ from a set of data.

9. Show that for the distribution

$$S(p) = p^\alpha - (1 - p)^\beta$$

the probability-weighted moments are

$$\omega_{r,s} = B(r + \alpha + 1, s + 1) - B(r + 1, s + \beta + 1),$$

where $B(.,.)$ are beta functions. Use this result to re-do
Problem 6 using the method of probability-weighted
moments.

10. The calculation of $p_{(r)}$ in Problem 6 gives a set of values
that can be regarded as the set of ordered values from a
uniform distribution that would exactly simulate the sample using the fitted model. This fact could be used to provide alternative criteria of fit, using the $p_{(r)}$ as data.
Possibilities are $\Sigma[p_{(r)} - r/(n + 1)]^2$ or some statistic
designed to test uniformity. Re-examine Problem 6 on this
basis. [See the starship method of estimation, e.g., King
and MacGillivray (1999)].

11. Observations are obtained from an exponential distribution with unknown threshold λ.

$$Q(p) = \lambda - \eta \ln(1 - p).$$

A matching is carried out based on the median of the data, m, and using $x_{(1)}$ as the natural estimate of the median of the distribution of the smallest observation. Show that λ is estimated by

$$\hat{\lambda} = (nx_{(1)} - m)/(n - 1).$$

12. The following frequency table gives the results of some industrial measurements. Fit, using the method of distributional least absolutes, the following distributions to the data: (a) skew logistic; (b) generalized lambda; (c) four-parameter lambda; and (d) five-parameter lambda. Compare the fits obtained.

$250^+ - 260$	3
$260^+ - 270$	5
$270^+ - 280$	14
$280^+ - 290$	34
$290^+ - 300$	31
$300^+ - 310$	27
$310^+ - 320$	15
$320^+ - 330$	9
$330^+ - 340$	10
$340^+ - 350$	5
$350^+ - 360$	3
$360^+ - 370$	3
$370^+ - 380$	0
$380^+ - 390$	2
Total	161

13. The data below are a set of failure time data. Fit by the method of distributional least absolutes using the following distributions: (a) The Weibull; (b) the Weibull with $\beta = 1$, i.e., the exponential; (c) the power $\times$ Pareto; and (d) the uniform $\times$ Pareto (i.e., (c) with $\alpha = 1$). Compare the fits using the least absolutes criterion and also the correlation between fit and observation.
Data: 35, 58, 66, 83, 84, 91, 97, 104, 104, 108, 111, 136, 137, 137, 138, 168, 186, 197, 211, 212, 256, 257, 346.

CHAPTER 10

Validation

10.1 Introduction

Validation is the process of deciding whether an identified and fitted model is indeed valid for the data and situation under consideration. There are several situations where validation becomes important:

(a) Where the identification process provides several possible models that have then been fitted, a choice between them needs to be made using more precise techniques than those used at the identification stage.

(b) Where a model with prespecified parameters is recommended for your data by an outside source.

(c) Where a well-used and trusted model is to be used in a new situation.

In case (a), where one is using one's own data and fitted models, care needs to be taken with the data used for the validation process. The general recommendation is that the validation should be carried out using different data from that used for the fitting. The reason for this is simply that the parameters will have been chosen to give the best fit, against some criteria, for that specific set of data. Such a good fit may not be achieved with further sets of data when the model is refitted and will almost certainly not be achieved if the initial fitted model is used unaltered. Having said this it is usual to have an initial look at the validity of the model against the data used for fitting. If the fit looks poor for this data, it will almost certainly be worse for any other data. A simple process of validation with large data sets is to use part of the data for identification and fitting and the remainder for validation. We refer to the data used for validation as the **validation data**. One way of using all the data in the validation process is to split the data in half. Use the first half for fitting and the second for validation and then repeat with the halves interchanged. This is called **cross validation**. Given sufficiently powerful software all the data save one observation could be used for fitting and the removed observation compared with its predicted value, $\hat{Q}(p_r^*)$, within the full sample. This is then repeated n times so that all the data is tested against predictions from the rest.

There are many aspects of validation, but here we concentrate only on those of direct relevance to modelling with quantiles.

10.2   Visual validation

Q-Q plots

The first question to address after fitting a model is, "Does the fitted model look right?" The natural plot to start with is the fit-observation diagram or Q-Q plot of the validation data against the corresponding fitted values. The fitted values may be based on the rankits, if these are available, or the median rankits. One possibility that the approach of median rankits provides is that the ordered observation can be compared, not only with its median value M_r, but also with some other quantiles. For example, the plot can give $x_{(r)}$ with the central 95% or 99% interval given by the fitted quantiles $\hat{Q}(p)$. These come directly from, for example,

$$\hat{Q}(\text{BETAINV}(0.025, r, n + 1 - r)), \quad \hat{Q}(\text{BETAINV}(0.975, r, n + 1 - r)).$$

Figure 10.1 illustrates this approach.

Another means of clarifying the Q-Q plot is to modify it by the addition of lines at p, $1 - p$, $\hat{Q}(p)$ and $\hat{Q}(1 - p)$ for $p = 1/4$, $1/8$ and $1/16$. These form a set of boxes to which the fitted median line is added. This is Parzen's quantile box plot. With the right model the proportions of the data lying in the three boxes are approximately 0.5, 0.75 and 0.875.

Density probability plots

For the fitted model the shape of the distribution is shown by $\hat{f}_p(p_r^*)$ plotted against $\hat{Q}(p_r^*)$, using the median rankit probabilities, p_r^*. If, however, on the same plot we show $\hat{f}_p(p_{(r)})$ against $x_{(r)}$ where $x_{(r)} = \hat{Q}(p_{(r)})$, then we have both observed shape and fitted shape shown on

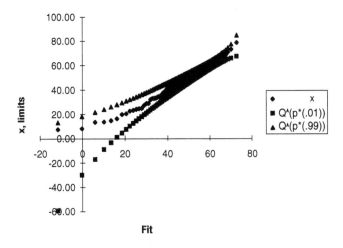

Figure 10.1. A fit-observation plot with 98% limits

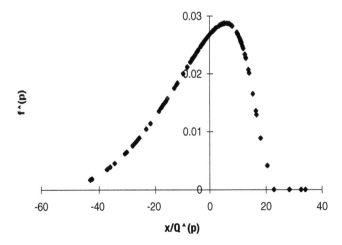

Figure 10.2. A density probability plot

the same plot. For situations where, for example, there is skewness in the data and a symmetric distribution is fitted, then this plot clearly shows the nature of the deviation from the model. Figure 10.2 illustrates the surprises one sometimes gets in plotting. The three values correspond to x and $Q(p)$ with very low likelihood. A closer inspection of the data revealed that one observation was an outlier, and the other two points were due to insufficient iterations of the algorithm for $p_{(r)}$. A paper by Jones and Daly (1995) gives a discussion of these plots, which they call **density probability plots**.

Residual plots

The Q-Q plot suffers from two disadvantages. First, the line of interest is sloping. Second, the natural variability of the points about the line is not constant, being greater at the ends of the data. The latter problem is solved by the introduction of the additional lines in the Q-Q plot. An alternative is to use the distributional residual of an observation, $x_{(r)}$, relative to its fitted value, fitx, which is simply

$$\text{residual} = x - \text{fitx} = x_{(r)} - \hat{Q}(p_r^*).$$

For years statisticians have been standardizing the residual, using the normal distribution as a means of looking at the fit of linear models. For this the values of ± 2 correspond approximately to the 2.5 and 97.5 percentiles. A corresponding process for a general distribution is to use the **standard distributional residual**, se_r, given by

$$se_r = e_r/[0.5\{\hat{Q}\,(\text{BETAINV}(0.975,\ r,\ n + 1 - r))$$
$$- \hat{Q}\,(\text{BETAINV}(0.025,\ r,\ n + 1 - r))\}].$$

For this we would expect around 5% of observed standard residuals with values numerically greater than one. For strongly skewed models or those constructed with differing tails, it may be useful to separate the residual analysis of the two tails. Suppose that $\hat{UD}\,(0.975)$ is the fitted upper 2.5% difference. We can then standardize the residuals to the right/left of the sample median using, for example, $se_{u,\ r} = e_r/[\hat{UD}\,(0.975) - \hat{M}_r]$ for r greater than the mid-value. Plots of these standard residuals can then be made against p and x to give a general feel for the quality of fit. Note that although standardization deals with the changes in variability of the order statistics the high correlation is still a problem. The distinction between similarly shaped distributions may be hidden by this feature, unless adequate data are available. This caution points again to the need to treat our models not as the truth, but as representations that pick up the main features of the population.

Further plots

A wide range of different population and sample plots were introduced in previous chapters. A comparison of any of the sample plots with the corresponding population plots for the fitted model will provide useful comparisons for identification. Table 3.3 gives the main population formulae and Tables 8.2 and 8.3 give the most useful sample equivalents.

Unit exponential spacing control chart

As we have already noted, a major problem in interpreting the fit-observation plots is that they are based on the ordered observations, any one of which is inherently highly correlated to those around it. This correlation produces the snake-like shapes shown by these plots. It is sometimes not clear whether what one is seeing is simply the effect of the correlation or is the result of some systematic deviation of the "true" model from the fitted one. It is therefore useful to create a form of plot where the plotted values are independent of each other. This can be achieved by bringing together two previous results. In Section 3.3 it was shown that the transformation

$$y = -\ln(1 - F(x))$$

transforms any variable x to a unit exponential variable y. This transformation gives a unique link. [It can be shown that for the exponential distribution multiples of the differences between successive exponential order statistics give values which are independent variables from a unit exponential distribution.] This again is a unique feature of the exponential distribution. These results provide for the construction of a plot of what may be termed **unit exponential spacings**. These are values independent of the unit exponential whose creation depends on having the correct fitted model. The various unique links in the argument imply that if the final data are not compatible with the unit exponential distribution, then the choice of $Q(p)$ must be wrong. The formal calculations are

(a) Calculate from the ordered data

$$y_{(r)} = -\ln(1 - F(x_{(r)})),$$

using the fitted $F(x)$, if this is explicit, or using the numerical method of Section 3.5 to obtain $p_{(r)}[= F(x_{(r)})]$ from $x_{(r)}$.

(b) Evaluate the corresponding independent unit exponentials from

$$v_{(r)} = (n + 1 - r)(y_{(r)} - y_{(r-1)}), \text{ with } y_{(0)} = 0.$$

(c) Plot these against r on the unit exponential spacing plot.

(d) Plot on the same graph one or two "control limits," W and A, as a help in interpretation. These are lines plotted at the 95% and 99% quantiles. They are obtained simply from

$$W = -\ln(1 - 0.95),$$

$$A = -\ln(1 - 0.99).$$

Figure 10.3 illustrates a plot based on 100 observations using a generalized lambda distribution, fitted by maximum likelihood, and showing just the 99% control limit. With 100 observations we might expect one or possibly two over this limit; however, the data appear well behaved except for one value. As this is not the end value it is not the effect of an outlier, but due to the relative positions of two near-end values. As often happens, the plot gives a generally positive validation but raises questions in the process. The fit-observation plot is used in parallel to supplement the information available.

10.3 Application validation

The process of identification is for the most part independent of the application for which the fitted model is required. It is therefore important at this stage to seek to validate the use of the fitted model against the requirements of the application.

As an initial consideration it may be possible to define specific criteria for the application, for example, cost functions that measure the financial penalties arising from the difference between the true and fitted models. These may sometimes be used as criteria of estimation, but this is not often possible or convenient. Thus these applications criteria can be used at this stage to get some feel for the value of the fitted model. Most estimation methods involve just one criterion, whereas in practice we may be interested in several. The values of these criteria may be derived using the validation data and fitted model.

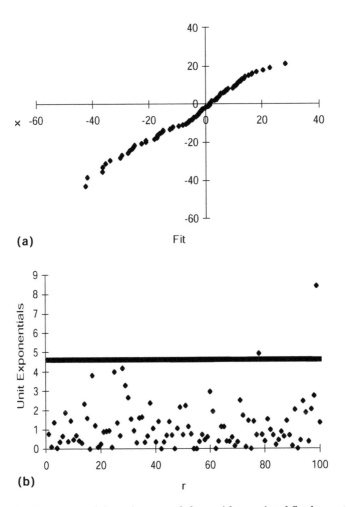

(a)

(b)

Figure 10.3. Unit exponential spacing control chart with associated fit-observation plot

In Section 4.4 we pointed out the ease with which quantile form models can be simulated. We are now in a position to use the fitted model to simulate data. This data can then be used in the way the application would use future data and the properties of the methodology explored.

Example 10.1: A routine statistical procedure was developed to select the significantly large individual values, from samples, e.g., those exceeding $\hat{Q}(0.99)$. This required the model to be re-estimated for each new, large data set. The generalized lambda distribution was used as a flexible model, well validated from past data. However, there will be errors due to variability in the estimates. Individuals will be selected when they are in fact below the population $Q(0.99)$ and conversely individuals will be missed who are above this value. The properties of such a technique can straightforwardly be explored using simulated data sets derived from some initial fitted model (and possibly models with designed deviations from the initial model). The proportion selected for the true model is 0.01. Samples can now be simulated from the initial model and the proportions actually selected, examined, and compared.

A further validation approach may emerge when particular features of the application area are studied. In some applications the tailweight of the distribution is important. In others, there are important functions related to $Q(p)$, for example, the hazard function of reliability studies, which we will discuss in Section 11.2. In these situations comparisons can be made between the validation data and fitted model based on such specific functions and features.

10.4   Numerical supplements to visual validation

Sometimes the look of two plots is very similar and it is useful to have some numerical measures to supplement the analysis. As the plotting of the fit-observation diagram is central to the validation of models, this is the natural place to look for convenient measures. Suitable measures of the quality of the fit are

> The correlation between fitted values and actual values.
> The Mean Square Distributional Residual, $\Sigma(x_{(r)} - E(X_r))^2/n$.
> The Mean Absolute Distributional Residual, $\Sigma |x_{(r)} - M_r|/n$.

There are no rules about the interpretation of the magnitudes of the second two; however, experience in any field of application will soon indicate the order of the values that are achievable and reasonable. They clearly provide a way of comparing alternative models.

10.5   Testing the model

We now turn to the formal testing of proposed models against validation data. There are a variety of methods available.

Goodness-of-fit tests

Suppose for simplicity we divide the p-axis into m equal sections using $p_j = j/m$, $j = 1, 2, ..., m - 1$; $p_0 = 0$, $p_m = 1$. If $w_j = \hat{Q}(p_j)$ and f_j is the number of observations in the new data set lying in the interval (w_{j-1}, w_j), then the expected value of f_j is n/m for all j. This fact is used to construct the test statistic C where

$$C = \Sigma[\{(f_j - n/m)^2\}/(n/m)].$$

As a general rule, statistics of the form

$$\Sigma[\{(\text{Observed} - \text{Expected})^2\}/(\text{Expected})]$$

have approximately a χ^2 distribution and in this case it has $m - 1$ degrees of freedom. If the new data is very different from the fitted model, the value of C will be larger than indicated by a χ^2 distribution. The χ^2 distribution is well tabulated and available in software as CHIINV(probability, degrees of freedom) or some such function, Thus we generate from the χ^2 quantile function, the value of, say CHIINV(1 − 0.95, $m - 1$). If C exceeds this, then we must doubt the validity of our old model. A look at the individual terms, $f_j - n/m$, may indicate where the model is fitting badly. It should be noted that the approximation is poor if the expected frequency, n/m, is less than about five. This requires that $m \leq n/5$.

Testing using the uniform distribution

If we have a new set of data $x_{(r)}$ and the fitted $\hat{Q}(p)$, we can derive, as in Section 4.5, the corresponding set of $p_{(r)}$. If the model is valid, these will be a set of ordered variables from a uniform distribution. We can therefore test the validity of the model by testing the uniformity of the distribution of the p. There are a variety of such tests. One simple one is based on the concept of entropy. This is defined for a distribution $f(x)$ by

$$H(X) = \int_{DR} f(x)\ln f(x)dx$$

$$= \int_0^1 \ln q(p)dp.$$

For the uniform distribution, $H(X)$ has its minimum value of zero (since $q(p) = 1$) and it reaches its maximum value of $\ln[2e\pi\sigma^2]/2$ for the normal distribution. A very simple sample estimate of $H(X)$, using the sample estimate of $q(p)$ from Section 8.4, gives

$$\hat{H} = -\Sigma_0^n \ln[(p_{(r+1)} - p_{(r)})(n + 1)], \text{ where } p_{(0)} = 0 \text{ and } p_{(n+1)} = 1.$$

For large n and under the hypothesis of uniformity, it can be shown that the statistic

$$t = (\hat{H} - nE)/\sqrt{n},$$

where E is Euler's constant, $(E = 0.5772)$ has approximately a normal distribution $N(0, \pi^2/6 - 1)$. The value of t based on the data can thus be used to test the model.

Tests based on confidence intervals

The previous tests hypothesized knowledge of the complete model form and all parameters. Sometimes we may assume the model form but wish to test just one parameter. This would occur, for example, if we felt that it was just the position or scale that had changed from the original situation, but the form of the distribution and the other parameters had remained unchanged. In Section 9.7 the idea of an interval estimate of a parameter θ was introduced. This gave an interval (l, u), based on the data, that has, say, a 95% probability of enclosing the true value, θ_0. If the model specifies θ_0 as the value to be tested and the interval does not contain θ_0, then the hypothesis that $\theta = \theta_0$ is rejected at the 5% level of significance. This provides a simple testing procedure for any parameter for which an exact or approximate confidence interval can be constructed.

Tests based on the criteria of fit

Suppose that we can accept the form of the distribution but wish to test either the full set of k-specified parameters, θ_0, or just some subset of them, say just $\lambda = \lambda_0$, then the log-likelihood ratio $r(\theta_0)$ introduced in Section 9.7 provides the basis for testing. If for the validation data the maximum log likelihood is $\ell(\hat{\theta})$, then the quantity D, where

$$D = -2r(\underline{\theta}_0) = 2[\ell(\hat{\underline{\theta}}) - \ell(\underline{\theta}_0)],$$

has approximately a χ^2 distribution with k degrees of freedom. If the validation data is consistent with the specified parameters, then D will be small. If it is significantly large, then the specified parameters are rejected en bloc. If some of the parameters are specified, $\underline{\theta} = \underline{\theta}_0$, but there is doubt about one, say ϕ, for which the model specified ϕ_0, then a similar test statistic is

$$D = -2r(\phi_0, \underline{\theta}_0) = 2[\ell(\hat{\phi}, \underline{\theta}_0) - \ell(\phi_0, \underline{\theta}_0)],$$

where $\ell(\hat{\phi}, \underline{\theta}_0)$ is the maximum log-likelihood over ϕ fixing the other parameters at $\underline{\theta}_0$. This has approximately a χ^2 distribution with one degree of freedom.

It may be that we wish to test one (or a number of) parameter(s), without assuming anything about the others. In this case we can still use the χ^2 approximation in the form

$$D = -2r(\phi_0, \hat{\underline{\theta}}) = 2[\ell(\hat{\phi}, \hat{\underline{\theta}}) - \ell(\phi_0, \hat{\underline{\theta}}(\phi_0))]$$

where $\ell(\hat{\phi}, \hat{\underline{\theta}})$ is the log-likelihood at the overall maximum and $\ell(\phi_0, \hat{\underline{\theta}}(\phi_0))$ is the maximum of the log-likelihood with ϕ_0 fixed. The function of ϕ_0, $\ell(\phi_0, \hat{\underline{\theta}}(\phi_0))$, is called a **profile likelihood**. Note that all these quantities are obtained using the basic layout of Table 9.6 simply by setting appropriate initial values and choosing which parameters to use in the maximization. D has a χ^2 distribution with degrees of freedom equal to the number of ϕ_0 parameters.

Note that the above χ^2 criteria can be translated into values for the likelihood ratio. For example, with the complete model the test will be based on comparing the calculated $R(\theta)$ with

$$R_{\min}(\theta_0) = L(\underline{\theta}_0)/L(\hat{\underline{\theta}}) = \exp(-\chi^2/2).$$

Table 10.1 gives the values of $R_{\min}(\theta_0)$ corresponding to some significance levels of χ^2. If the observed ratio is less than the given values, doubt is expressed about the specified model or parameters at the given level of significance. Conversely, we accept the model as valid if $R \geq R_{\min}$. All these various tests are approximations needing large samples.

Degrees of freedom	Level of significance		
	10%	5%	1%
1	0.03625	0.01945	0.00445
2	0.0100	0.00500	0.00100
3	0.00344	0.00163	0.00029
4	0.00131	0.00059	0.00010
5	0.00053	0.00023	0.00004

Table 10.1. Significant values of the likelihood ratio, R_{min}

The ratio R can be regarded as the criterion for validating the specified model using maximum likelihood as the method. We could equally use the criterion originally regarded as appropriate for fitting the model. For the methods of DLS and DLA we could define ratios as

1.	Define distribution	$Q(p;\underline{\theta})$
2.	Derive quantile density	$q(p;\underline{\theta})$
3.	Set parameters from specified model	$\underline{\theta}_0$
4.	Set median-p_r, $\mu_{(r)}$, or initial $p_{(r)}$	$p_r^* = $ BETAINV$(0.5, r, n + 1 - r)$, $\mu_{(r)}$ from formulae or tables, $p_{(r),0} = r/(n + 1)$
5.	Derive $p_{(r)}$ (where $y_{(r)} = Q(p_{(r)};\underline{\theta}_0)$)	iterate for each r $p_{(r)new} = p_{(r)} + (y_{(r)} - Q(p_{(r)};\underline{\theta}_0))/q(p_{(r)};\underline{\theta}_0)$, see Section 4.5(b)
6.	Calculate criterion	$C(\text{data } y_{(r)}| p_{(r)}; \underline{\theta}_0) = C(\underline{\theta}_0)$
7.	Search for $\underline{\hat{\theta}}$ to optimize 6 (involving iterating steps 3 to 6, possibly keeping some parameters at specified values)	$\underline{\theta} = \underline{\hat{\theta}}$, optimum $C = C(\underline{\hat{\theta}})$
8.	Fitted distribution	$\hat{Q}(p;\underline{\hat{\theta}})$
9.	Fitted quantile density, if needed	$\hat{q}(p;\underline{\hat{\theta}})$
10.	Calculate criterion ratio	$R = C_{DL}(\underline{\hat{\theta}})/C_{DL}(\underline{\theta}_0)$ or $L(\underline{\theta}_0)/L(\underline{\hat{\theta}})$
11.	Validate	valid if R greater than R_{min} (Table 10.1 for maximum likelihood or empirically chosen values for other criteria)

Table 10.2. Validating algorithm for least/maximum methods (y = validation data)

$$R = C_{DLS}(\hat{\theta})/C_{DLS}(\theta_0) \text{ or } R = C_{DLA}(\hat{\theta})/C_{DLA}(\theta_0)$$

where as before the estimates come from the validation data and doubt is cast on the specified parameters if the ratio is too small. In this case, however, we cannot easily associate levels of significance with the values. The procedure for looking at only one or two of the parameters parallels that using the likelihood ratio, with parameters other than those of interest kept at their specified values. It will be seen that the methodology is an extension of the original method of fitting the distribution by the least or maximum methods as shown in Table 9.5. We modify this table in Table 10.2 to show the overall approach.

10.6 Problems

1. It is assumed that $Q(p) = \lambda + \eta S(p)$, where $S(p)$ and η are known. Show that an approximate 95% test of $\lambda = \lambda_0$ can be obtained by rejecting this hypothesis if the sample median lies outside the interval

   $$[M - 1.96 * 2\eta/\sqrt{n}, M + 1.96 * 2\eta/\sqrt{n}].$$

2. The Rayleigh distribution is a Weibull with shape parameter 1/2,

   $$Q(p) = \eta\sqrt{[-\ln(1 - p)]}.$$

   Show that, corresponding to the previous example, the median-based test for the hypothesis $\eta = \eta_0$ is based on the interval

   $$[\eta_0\sqrt{\ln 2} - 1.96\eta_0/\{2\sqrt{(n \ln 2)}\}, \eta_0\sqrt{\ln 2} + 1.96\,\eta_0/\{2\sqrt{(n \ln 2)}\}].$$

3. Suggest how the generalized Pareto can provide a test for the exponential distribution against clearly defined alternatives.

4. A number of models were fitted to data in Problem 9.1. Reconsider this data as a validation exercise against the fitted models. Simulate data from one of the fitted models and validate it against one of the other fitted models.

5. Problem 9.5 fitted data using percentile-based methods with the shape parameter depending on the right half of the sample. Validate the fitted model against the left half of the sample.

6. Problem 9.1 gave a generalized lambda fitted to some data. Section 4.5(c) gave an iterative method of finding $p_{(r)}$, where $x_{(r)} = \hat{Q}\,(p_{(r)})$. For a valid model the $p_{(r)}$ are from a uniform distribution. Derive the $p_{(r)}$ for the generalized lambda fit and plot against the median rankits for the uniform distribution (which are the median-p, $p_r^{*} = $ BETA-INV$(0.5, r, n + 1 - r)$). Comment on the plot obtained. This form of visual validation is a variant of the "p-p plot," which plots $p_{(r)}$ against the mean rankits, $r/(n + 1)$.

CHAPTER 11

Applications

11.1 Introduction

The test of all methods lies in their practical use. Indeed, the operations of statistical modelling cannot be divorced from the specific
requirements of the application. To deal seriously with applications
would require a separate book. The purpose of this chapter is to give
some snippets that illustrate how quantile-based methods can be of
relevance to areas of application.

11.2 Reliability

Definitions

Reliability theory is the study of the probabilities of items or systems
of items working. Indeed, the formal definition is

$$\text{Reliability} = \text{Prob(item or system works).}$$

The item may be a component, a machine, a process, or even a person.
The area of specific relevance to the topics of this book concerns
situations where reliability is a function of time. The item works when
first installed or switched on but at some time, t, it fails and ceases
to work. There are three interrelated descriptions possible of such
situations. These are based on three statistical functions:

(a) The time to failure distribution. The quantity t can be
regarded as the observed value of the random variable,
T, having some distribution $f(t)$. Many different distributions have found practical use as **failure time distributions**. Of particular importance are the exponential and
the Weibull.

(b) The Survivor Function. The reliability at time t depends on whether the item survives at least to time t. The probability of this is

$$\text{Prob}(T \geq t) = S(t) = 1 - F(t) = 1 - p = q,$$

which is called the **survivor function**. The quantile function for the time at which the reliability becomes q is thus $Q(1 - q)$.

(c) The Hazard Function. The **hazard function** is defined as the probability of immediate failure at time t. Thus the item fails in a small time, dt at t, conditionally on working up to time t. The rule of conditional probability thus gives

$$h(t)dt = \text{Prob}(\text{failure in } dt | \text{survives to } t)$$
$$= f(t)dt/S(t).$$

p-Hazards

To convert $h(t)$ from a function of t to one of p it can be noted that $S(t)$ is just $1 - p$ and $f(x)$ becomes $f_p(p)$. Thus we have a **p-hazard function** defined by

$$h_p(p) = f_p(p)/(1-p)$$
$$= 1/[(1-p)q(p)].$$

In emphasizing the role of quantiles in reliability the p-hazard provides the natural tool. The following provides some examples:

(a) The Exponential Distribution. For the exponential distribution with parameter η, the p-hazard function is

$$h_p(p) = 1/\eta.$$

Thus the instantaneous failure probabilities are constant. This is a consequence of the relation of the exponential to the Poisson process. The Poisson process is, roughly, any series of events where there is a constant probability of events occurring and where the occurrence of one event has no effect on the possibility of other events occurring

after it. In such a process the exponential is the distribution of times between events. A feature of the indepen-
dence of events in such a process is that there is no
'memory.' The fact that a component has already lasted a
very long time has no effect on how long it is likely to last
in the future. Although this may not seem very appropriate in general, it will be appropriate where failures are
due to some form of 'accident' that can be reasonably
modelled by a Poisson process. It will be seen in what
follows that the constant hazard acts as a special case for
several other p-hazard functions.

(b)   The Weibull Distribution:

$$h_p(p) = (1/\beta\eta)[-\ln(1 - p)]^{(1 - \beta)}.$$

It is evident from the power of the term [....], which is an
increasing function from zero at $p = 0$, that the curve is
increasing for $\beta < 1$, constant for $\beta = 1$ and decreasing for
$\beta > 1$. An increasing function is called positive ageing, the
normal form, and a decreasing function is negative ageing.
Notice that $h_p(p)$ when converted back to $h(x)$ gives the
function of x as $x^{(1 - \beta)/\beta}$. For $\beta = 1/2$ this is just x, so the
hazard function $h(x)$ is linear in x. This special case distribution is called the **Rayleigh distribution**. For $\beta = 1/3$
the hazard gives a multiple of x^2. Thus powers in x are
hazards for the Weibull (see Figure 11.1(a)).

(c)   The generalized Pareto distribution:

$$h_p(p) = (1 - p)^\beta.$$

Here we have positive, zero and negative ageing according
to the positive, zero or negative sign of β.

(d)   The symmetric logistic distribution:

$$h_p(p) = p/\eta.$$

Here we have the simplest increasing ageing, linear in p.

All the above p-hazards are constant, decreasing or increasing
functions. Items get better with use or they burn out with use. Sometimes, as with human beings, both processes occur together. There is

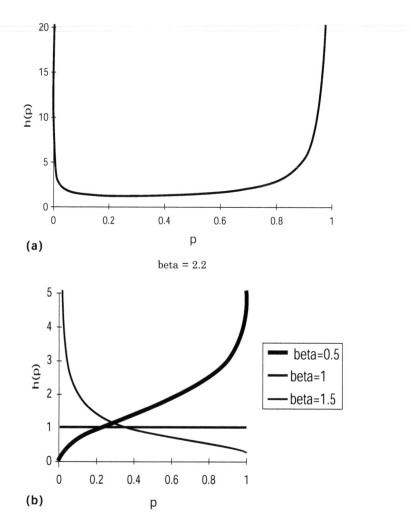

Figure 11.1. The p-hazard functions for (a) Weibull and (b) Govindarajulu distributions

a 'burn-in' period of high initial failure rates, followed by a fairly steady working life, followed by a final 'burn-out' with age. The shape of the hazard function in these cases leads to their natural description as 'bath-tub' hazards. If we use quantile modelling to look for such curves, a natural starting point would be p-hazards of the form

$$h_p(p) = 1/[\eta p^\alpha (1-p)^\beta], \ \alpha, \ \beta > 0,$$

which increases as p approaches zero or one. For this p-hazard, the quantile density takes the form $\eta p^{\alpha}(1 - p)^{\beta - 1}$. A look back at Example 6.6 shows that the Govindarajulu distribution is a case of this general form. Figure 11.1(b) shows the form of the p-hazard for a Govindarajulu distribution.

As a final note, it should be observed that the p-hazard provides a means of identifying distributions. The sample-based $\tilde{q}\,(p_r)$ can be obtained from the data, p_r from the median probability, p_r^* and hence $\tilde{h}\,(p_r^*)$ are obtained from a table. It is then plotted against p_r^* and the plot compared with standard p-hazards, such as those discussed above.

11.3   Hydrology

Hydrology is concerned with matters relating to rainfall, river flow, etc. The provision of fresh water requires knowledge of the chances of drought, the design of reservoirs, and the chances of high rainfalls for the design of flood drains and runoffs. The modelling of tails of distributions and the knowledge of extreme quantiles are thus central to the statistics of hydrology.

Suppose $Q(p)$ is the quantile function for a hydrological variable, such as annual minimum rainfall or annual maximum flood height, then for some given high p_0, say 0.99, there is a $100(1 - p_0)\%$ chance of an **exceedence** beyond $Q(p_0)$. The probability of the first exceedence to occur in year k is

$$\text{Prob}(K = k) = p_0^{k - 1}(1 - p_0).$$

This is the geometric distribution mentioned previously. The mean value for k is $1/(1 - p_0)$. This mean is called the **return period**, T, for a $(1 - p_0)\%$ exceedence event. Looking at the situation for a given T, for example, if T is 100 years, then $Q(0.99)$ will be the value of the 100-year flood.

The importance of the extreme tails in hydrology and the occurrence of very long-tailed distributions in hydrological data have led to an emphasis on distributions that model such data and provide simple analysis of the quantiles. Thus the literature of hydrology frequently presents distributions in terms of the quantile function, although referred to just as the inverse of the CDF. There have in fact been distributions defined and explored first by the quantile function rather than the CDF or PDF; for example, the **Wakeby distribution** (see Houghton (1978)) in canonical form is defined by

$$Q(p) = \lambda + (\eta/\alpha)[1 - (1 - p)^\alpha] - (\nu/\beta)[1 - (1 - p)^{-\beta}].$$

It will be seen that the form of the distribution is the sum of two generalized Pareto distributions with parameters of different signs in the natural form, although β may be made negative. Being a flexible five-parameter model this has been shown to give an excellent model of flood data. The distributional range is (λ, ∞) if all parameters are positive or zero. If β alone is negative, then the range is $[\lambda, \lambda + (\eta/\alpha) - (\nu/\beta)]$.

When discussing the method of moments for estimation it was observed that where there are several parameters needing higher moments to be calculated, the method leads to highly variable estimators. This applies to the highly parameterized models of hydrology. The situation is made worse by the fact that the long-tailed situations naturally have data sets containing extreme observations, which dominate the estimates of skewness and kurtosis. Thus it is natural to look for methods of estimation that are less sensitive and possibly are linear in the data. It is not surprising that methods like the method of probability-weighted moments were largely developed in the context of hydrology. The context of estimation almost always raises issues about the methods used. Methods suitable in one application may not be suitable in another. Indeed the application often provides a criterion of fit specific for the situation. This may not lead to a formal method of estimation but it can be used to provide guidance as to which of the standard methods is appropriate.

Example 11.1: A distribution, $Q(p)$, is fitted to give estimated quantiles as $\hat{Q}(p)$. If this is used in the design of, say, a flood defence, then the estimation errors $e_p = Q(p) - \hat{Q}(p)$ will lead to errors in the design of the defences. If for the p of a given return period e is positive, we will have underdesigned the defences so that, for example, the 100-year flood defence will be breached more frequently than the average 100 years. The costs of the errors can thus be defined as some simple function of e_p. Different methods of estimating the distributional parameters can then be compared by simulation, using $Q(p)$ in the simulation and then fitting to get $\hat{Q}(p)$. Fitted models are often used as the bases for elaborate simulations of designed water storage facilities, etc. In this case we require the distributional fit to be good over all values of p. A criterion in this situation could be $\int E(e_p^2)\,dp$ or its experimental value from the simulations.

Example 11.2: In areas where long tails and extreme values are common, the study of right tails is often carried out by introducing a high threshold and using only data above the threshold value for fitting. This is a rather extreme approach. Our previous studies have suggested fitting with discounting (Section 9.5) and using more flexible models that have different forms in the two tails (Section 6.3).

11.4 Statistical process control

Introduction

In these days of emphasis on quality the use of statistical methods to control production processes (termed **statistical process control** or **SPC**) has grown rapidly. An almost traditional assumption of the SPC techniques used in many organizations has been that of the normality of the data. For example, the author visited a factory where there is an audible warning whenever a process gets 'out of control.' This warning seemed a regular part of the firm's background noise. The SPC methods used to monitor the process and generate the warnings were based on assuming normality. Much of the data generated by measurements on the processes used showed considerable non-normality. It was therefore not always a loss of control that was the problem; the wrong assumption also added to the noise. A particular problem in many organizations is that their processes actually produce data with long-tailed distributions. Where this is correctly identified significant quality improvement can be made by working on the process to reduce the tailweight. The assumption of normality is widespread in the methods used in quality studies. The traditional approach to non-normality has been to transform the data to normality (see, for example, Chou, Polansky and Mason (1998)). A contribution to SPC by the use of quantile methods can be found by adopting SPC to cover a much broader class of distributions than has been traditional. We illustrate briefly.

Capability

When things are manufactured or services delivered, the customer often specifies precisely what is required: some level of accuracy of manufacture or some maximum delivery time. Sometimes the providers set themselves such targets or specifications. The question then arises as to whether or not the provider is capable of meeting the specification. Can the machines manufacture to that level of precision? Can some guarantee be provided to the customer? In a random world the provider will never be able to guarantee that the process will never produce a result out of specification. To be realistic, the language of probability is needed. Many people, however, are still uncomfortable with such language, so historically an approach was developed that depended on comparisons and the use of simple numerical indices. It also started with an assumed normality. Customer specifications may be one-sided, such as a requirement for a job to be done in less than a specified time, or two-sided, such as the limits on some dimension of a part. Suppose in the latter case the limits are (L, U), then $U - L$ is the **specification range**. This range may or may not be compatible with the inherent variability of the underlying process. Suppose this variability is investigated on the basis of a sample of, say, 100 observations. This is often referred to as a **capability study**. An outcome of the study will be a model, $\hat{Q}(p)$, based on the data and having fitted parameters. The **capability range**, CR, for this model is the range of values that will contain almost all likely data. Traditionally this has been based on $[\hat{Q}(0.00135), \hat{Q}(0.99865)]$. The definition is thus

$$CR = \hat{Q}(0.99865) - \hat{Q}(0.00135)$$

These values are used since for the normal distribution, which has been commonly assumed as well as frequently identified, they correspond to $[\mu - 3\sigma, \mu + 3\sigma]$, so $CR = 6\sigma$. For situations where the distribution is centred around the target value, a **capability index**, C_p, is defined by

$$C_p = \text{Specification Range/Capability Range},$$
$$C_p = [U - L]/\{\hat{Q}(0.99865) - \hat{Q}(0.00135)\}.$$

This index should be greater than one for a process to be regarded as capable. Owing to the potential misleading effect of the estimation involved in obtaining $\hat{Q}(p)$, most organizations look for a value above 1.3.

One problem arises if the mean, or in our version, the median, is not the centre of the specification. In this case each tail of the distribution is examined separately and the worst case used as the index. Thus we have the C_{pk} index.
If

$$C_U = (U - Q(0.5))/[\hat{Q}(0.99865) - \hat{Q}(0.5)],$$

and

$$C_L = (Q(0.5) - L)/[\hat{Q}(0.5) - \hat{Q}(0.00135)],$$

$$C_{pk} = \text{minimum}\{C_L, C_U\}.$$

A measure of the "off-centredness" of the process, when m is the mid-value of the specification, is

$$K = |Q(0.5) - m|/(U - m).$$

A little algebra will show that $C_{pk} = C_p(1 - K)$. Thus C_{pk} merges the effect of both the lack of precision and the off-centredness.

There are many variants of capability indices (see, for example, Kotz and Johnson (1993) and Gilchrist (1993)). The point of this brief discussion is just to underline, once again, that the quantile function provides the natural language to discuss the problems involved. It provides for a natural generalisation from the usually assumed normal distribution.

Control charts

In an ideal factory there would be, at least in some situations, perfect uniformity. All items produced on the machine would be identical; all items ordered by 4:00 P.M. would be delivered at 10:00 A.M. the next day. In reality, that is not the case. The causes of this non-uniformity can be roughly divided into **special causes**, which are those due to some specific and assignable factor in the situation, an operator error or a lorry accident, and **common causes**, which are the thousand-and-one chance small variations that lead to a statistical distribution for any variable in the processes. A major question is obviously how one distinguishes between the two.

This was first answered by Shewhart in 1929 through his use of **control charts**. Observations are made on some variable in a process, possibly a derived variable such as an average, which is assumed (on the basis of a capability study) to have a distribution $Q(p)$. The observed values are plotted on the chart. Also on the chart two types of lines are drawn. For a two-tailed distribution, action lines are drawn at $Q(p_A)$ and $Q(1 - p_A)$ and warning lines at $Q(p_W)$ and $Q(1 - p_w)$. If the underlying distribution is normal, the probabilities p_A and p_W are usually set at 0.00135 and 0.0228, which put the action and warning lines at $\mu \pm 3\sigma$ and $\mu \pm 2\sigma$, respectively. If an observation falls outside the action lines, then it represents an event of sufficiently low probability that it is taken to be due to a special cause. A search is made to identify that cause and then to take action to put it right and to prevent it happening again. A sequence of observations outside a warning limit will also suggest a special cause. In many charts these days the warning lines are not used. The variability that will be observed in the data between the action lines is that due to the common causes. The philosophy of modern quality improvement approaches is that the processes generating the data are constantly worked upon to steadily remove special causes and reduce the variability generated by common causes. Thus over time the charts should be revised to have narrower limits. Notice that even if all possible special causes were removed there would still be calls to action with probability $2p_A$, which is equivalent to one false alarm in every $1/2p_A$ values.

The conventional assumption of normality is not a necessity and any appropriate $Q(p)$ could be used in the chart, with the parameters fitted during the capability study.

Example 11.3: To illustrate the ease of use of the quantile model, consider a situation where small is best. For example, we have to get to the repair job as quickly as possible, or the amount of impurity has to be as small as possible. Let x be the measured variable and suppose it has an exponential distribution with specified scale parameter η. We take not just one observation but a sample of n observations and plot them on a control chart. The action line is at some value $x = a$ and action is taken if any observation lies above this line. If all the points are below the action line, then the largest must be. Thus we can simply concentrate on the largest value which, as was shown in Section 4.2, has distribution $Q(p^{1/n})$. If p_a is the probability of the chart correctly indicating that the process is in control, then we have $Q(p_a^{1/n}) = a$. For the exponential this leads to

$$a = Q(p_a^{1/n}|n) = -\eta \ln(1 - p_a^{1/n}).$$

Thus a can be found from the given (target) η and p_a set at some suitable high value such as 0.999 or 0.9973 to correspond to the standard normal control chart probability. Keeping to the former, the probability of action being taken when all is in control is 0.001. The **average run length**, ARL, between such false alarms is $1/(1 - p_a) = 1000$. Suppose the scale parameter suddenly increases to a value of $k\eta$, $k > 1$. The probability of exceeding the control limit is now p_k, where

$$a = -k\eta \ln(1 - p_k^{1/n}), \text{ but also } a = -\eta \ln(1 - p_a^{1/n}),$$

hence

$$\ln(1 - p_k^{1/n}) = \ln[(1 - p_a^{1/n})^{1/k}],$$

and

$$p_k = [1 - (1 - p_a^{1/n})^{1/k}]^n.$$

The ARL can thus be obtained from ARL $= 1/(1 - p_k)$ as a function of k. The formula can also be used for the choice of n given a required ARL for some specified k.

11.5   Problems

1.   The life of an item of equipment is distributed with quantile function $Q(p)$. Suppose it is known that the equipment was operational at time t_0, where $t_0 = Q(p_0)$. Denote the remaining, residual life by s with quantile function $Q_s(p)$. Show by using the methods of Section 6.8 that

$$Q_s(p) = Q[p(1 - p_0) + p_0] - t_0,$$

and derive the distributional range and median, by way of a check.

2.   Calculate the shape of the p-hazard for the following distributions:
(a) $Q(p) = p/(1 - p)$,
(b) $Q(p) = p^2(3 - 2p)$,
(c) $Q(p) = -(1 - p)^\beta(1 + \beta p)$.

3.   An electrical item has a life that is usually modelled by a Weibull distribution with shape parameter β. In a particular use there is a probability P that a fitting for the item fails when the circuit is switched on. Show that the model for the item and fitting has a distribution that corresponds

to a power Q-transformation of an exponential distribution with non-zero threshold.

4. A sample of a new product is tested to distruction and 18 observed lifetimes are available. The model assumed initially is the exponential distribution. Consider how to estimate the parameter η: (i) using distributional least absolutes; (ii) using maximum likelihood, for the situations where
    (a) The 18 observations were the complete sample.
    (b) The 18 observations were from a batch of 100, but were known to be all those that had failed in under 20 months.
    (c) The sample consisted of 25 items but it was decided at the beginning of the trial to stop the trial after the 18th item had failed.

5. Show that for the Weibull distribution a plot of the $\ln(p\text{-hazard})$ against $\ln[Q(p)]$ is linear. Examine how this might be used for identification of Weibull and exponential models. Apply your suggested method to the flood data of Chapter 1.

6. A component has a time to failure, t, with an exponential distribution, with $QF - \eta\ln(1 - p)$. The component is still operating at time t_0, with corresponding probability p_0. Show that the conditional remaining life of the component, $y = t - t_0$, is also $-\eta\ln(1 - p)$; i.e., there is no memory of the fact that the component has already been operating for time t. This no memory property is a unique feature of the exponential distribution.

7. In a "parallel system" of components the system works as long as one component works. Such a system has m components, each of which has a Weibull failure time distribution, $\eta[-\ln(1 - p)]^\beta$. Show that the system's time to failure is an exponentiated Weibull with quantile function $\eta[-\ln(1 - p^{1/m})]^\beta$.

8. In a "weakest link system" of components, the system fails on the earliest failure. Such a system has m components, each of which has an exponential failure time distribution.

Show that the system's time to failure is also an exponential distribution, but with scale parameter η/m.

9. A distribution of some hydrological use is the three-parameter log-normal with quantile function

$$Q(p) = \lambda + \exp[\mu + \sigma N(p)].$$

Show that this may be fitted explicitly using the method of percentiles, with the median, largest and smallest observations; for example, $\hat{\lambda} = [x_{(1)}x_{(n)} - m^2]/[x_{(1)} + x_{(n)} - 2m]$.

10. A capability study of a process assumed normality and found the mean to be on specification at 1000 and the standard deviation to be 1.4. The calculated capability was 1.17, which was judged acceptable. A further study showed the process to have a logistic distribution. Show that the capability should have been calculated at 0.97.

11. A control chart based on three observations is constructed for the scale parameter, η, of a logistic distribution. Upper and lower limits, $\lambda - u$ and $\lambda + u$, are constructed to give a small probability, p_0, of observations lying outside, i.e., $u = \eta \ln(p_0/(1 - p_0))$. Action is taken if the three observations simultaneously fall in each of the three regions of the chart. Show that the probability of this occurring is approximately $6p_0^2$. The scale parameter increases to $k\eta$. Find an expression for the average run length as a function of k.

12. A control chart for the scale parameter, η, of a Weibull distribution is to be devised for the situation where small is beautiful. Action is taken if the largest observation of a sample of n exceeds the single upper control limit. If the average run length is set at 1000 for the normal situation, derive an expression for the average run length for the situation where the scale parameter increases to $k\eta$.

CHAPTER 12

Regression Quantile Models

12.1 Approaches to regression modelling

The first columns of Table 12.4 show some data based on map measurements of distances between randomly chosen pairs of points in Sheffield (see Gilchrist (1984)). The straight line distance is measured with a ruler (the crow flies distance, x) and the shortest distance by road is also measured on the map (the car distance, y). The objective is to find a simple relationship between the two so that y can be predicted from x. The units were measured in cm on a 1:25000 map, i.e., the unit was a quarter of 1 km. Several features follow from a general conceptual modelling argument:

(a) When $x = 0$ then $y = 0$.
(b) The model must give $y \geq x$.
(c) In general we will have y increasing with x.
(d) A linear relation is intuitively a sensible first model, as we would expect to double y by doubling x.

The sample of 20 measured pairs (x_i, y_i) is plotted in Figure 12.1. On the basis of this a straight line looks to be a natural, simple starting model. Clearly there is a random error, e, around the deterministic line. The complete model as a first identification on both conceptual and empirical grounds is

$$y = \theta x + e; \ x \geq 0, \text{ and the slope } \theta \geq 1.$$

The above expression is a simple example of a **regression model** in which y is 'regressed' on x. The variable y is termed the **dependent variable** and x the **regressor variable**.

In more complex models there may be several regressor variables and they may appear in the equation in more complex ways, e.g., as x^2 or $\sin(2\pi x)$. Having decided on this initial model the next step is to

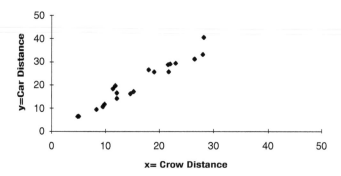

Figure 12.1. Crow–car data — scatter diagram

seek to fit it to the data. When the question of how to fit a model to a set of data was first raised by astronomers in the 18th century (who were looking at much more complex models), the natural approach adopted was to choose the values of the parameters to make the magnitudes of the errors overall as small as possible. Thus in this model they would have chosen the estimate $\hat{\theta}$ to make the sum of the observed absolute errors a minimum. The fitted value of y_i is $\hat{y}_i = \hat{\theta} x_i$, so their criteria would thus be

$$C_1 : \Sigma |y_i - \hat{\theta} x_i| \text{ minimum.}$$

This is the method of least absolutes, which was first introduced in the 1750s by Boscovitch (1757) (see Harter (1985)). Unfortunately it was found that this minimization was extremely difficult to carry out with the mathematical and computational tools of the day. At about this time calculus was developed and it was realized that if C_1, in our example was replaced by

$$C_2 : \Sigma (y_i - \hat{\theta} x_i)^2 \text{ minimum,}$$

then the minimization problems often had simple explicit solutions using calculus. Thus the method of least squares was developed and the solution for our problem is found by differentiating C_2 with respect to $\hat{\theta}$ and equating to zero to give

$$\hat{\theta} = \Sigma(x_i \, y_i)/\Sigma x_i^2 .$$

Table 12.1 shows the layout illustrated from the first half of the data. This gives an estimated slope of $\hat{\theta} = 1.29$.

$n = 20$; slope = 1.29; sigma = 2.38; average e = 0.06; Σe^2 = 107.30

r	Car distance y	Crow distance x	xy	fit	residual e	ordered e	p^*	normal fit
1	10.7	9.5	101.65	12.25	-1.55	-2.99	0.034	-4.27
2	6.5	5.0	32.5	6.45	0.05	-2.96	0.083	-3.24
3	29.4	23.0	676.20	29.65	-0.25	-2.53	0.131	-2.60
4	17.2	15.2	261.44	19.59	-2.39	-2.39	0.181	-2.11
5	18.4	11.4	209.76	14.70	3.70	-2.27	0.230	-1.70
6	19.7	11.8	232.46	15.21	4.49	-1.55	0.279	-1.33
7	16.6	12.1	200.86	15.60	1.00	-1.40	0.328	-1.00
8	29.0	22.0	638.00	28.36	0.64	-1.20	0.377	-0.68
9	40.5	28.2	1142.10	36.35	4.15	-0.93	0.426	-0.38
10	14.2	12.1	171.82	25.60	-1.40	-0.25	0.475	-0.08

etc.

Table 12.1. Least squares fitting — Crow–Car data

From the fitted line the **residuals**, e, can be calculated. These are the differences between the observed and the fitted values:

$$e_i = y_i - \hat{\theta} x_i.$$

A measure of the variability in the data about the fitted line is given by the sample variance of the residuals,

$$s^2 = \Sigma(e)^2/(n - 1).$$

If a normal distribution is assumed for e, then the residuals should be normally distributed with a mean of zero and a standard deviation estimated by s. From this a normal quantile plot of the residuals can be obtained, as in Figure 12.2. Notice that (a) this is a fit-observation plot for the stochastic element in the model and (b) the plot does not suggest a very good fit.

If we wish to use the criterion C_1 we need to apply numerical methods to find the estimated slope. A starting estimated value of the parameter, a sensible guess like $\hat{\theta}$ = 1.5, is used to produce **fitted values** $\hat{y}$. Residuals can then be calculated and the criteria $\Sigma|e_i|$ derived. The use of minimization procedures enables $\hat{\theta}$ to be chosen to find the slope that minimizes the numerical value of the criteria.

We have shown in the example that for the same regression model we can apply different criteria of goodness of fit. The method of least squares often leads to explicit formulae for the estimates and gives estimates that have good statistical properties. The method of least absolutes rarely gives explicit algebraic answers. A consequence is that

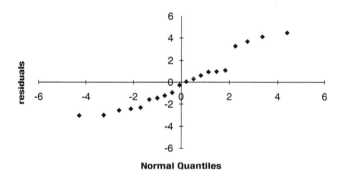

Normal Quantiles

Figure 12.2. Crow–car data — normal plot of residuals

least squares has been so commonly used that it is often forgotten that least squares is just one criterion out of a possible infinity. However, one must consider not just the statistics, but also the application. In our example the importance of an error of two miles is measured naturally as 2 and not as 2^2. Thus the reasonable criterion for this particular application is C_1 and not C_2. It should also be noted that both C_1 and C_2 imply a symmetry in the importance of positive and negative errors and consequently a usual assumption that they have symmetric, although often unspecified, distributions.

It is seen from the above that the estimates have been obtained with no specific reference to the distribution of e. The methods are thus referred to as **semiparametric** methods; only some of the model is parametrically defined. If a distribution is assumed, it is usually the normal with zero mean and a standard deviation which would be estimated by s above or a variant on it. Notice that the semiparametric approach focuses on the deterministic part of the model. If information is needed about the stochastic element the residuals are used to provide it in a totally separate exercise.

Another approach to regression is called **quantile regression**. As an example for the crow–car data one might have $y_p = \theta_p x + e$ as a linear model for the p quantile of y given x. As a more practical example, one might have the 99%, 95%, 75%, 50%, 25%, 5%, and 1% regression curves for the weight of male babies regressed on age. This set of regression curves would give a clear feel for potential weight problems of individual children given their age. In the light of previous discussions it is hardly surprising that $\theta_{0.5}$ can be found by the method of least absolutes. The fitted line in this case is called the **median regression** line. Median regression is illustrated in papers by Ying, Jung, and Wei (1995) and Jung (1996). To obtain a fitted p-regression, the criterion of C_1 is replaced by

$$\Sigma[p(y_i - \theta x_i)^+ + (1 - p)(\theta x_i - y_i)^+]$$

where $z^+ = \max(z, 0)$. Minimizing this criterion will give the fitted **regression p-quantile function** (see, for example, Koenker and Bassett (1978) and Bassett and Koenker (1982)). The papers by Koenker (1987) and D'Orey (1993) give computing algorithms for these procedures. Notice that these methods are also semiparametric and have to be used independently for each required p.

Our interest is in the use of quantile functions to fully model distributions, so it is natural to model the error term by $\eta S(p)$, where

$S(p)$ is a quantile function without any requirement of symmetry, and where η is the scale parameter. Putting the regression and quantile expressions together we get the p-quantile of y, given the specified x. For example, for the general crow–car model this gives

$$Q_y(p|x) = \theta x + \eta S(p).$$

This we could call the **regression quantile function** of y on x, sometimes called the conditional quantile function. The phraseology "of y on x" emphasizes that we are using x to predict y. The natural plot is of y against x. If it were vice versa, the model would need to be different to allow for a different form of error influencing the x rather than the y. Notice that our formulation shows both the deterministic and stochastic components of the model together with their parameters. This approach will enable us to estimate both regression and distributional parameters in one operation. The model is thus a full **parametric regression model**.

For illustration consider the Crow–Car problem to have normally distributed errors with zero mean and median and a standard deviation σ. The regression quantile function is thus

$$Q_y(p|x) = \theta x + \sigma N(p).$$

To fit this model initial values of both $\hat{\theta}$ and $\hat{\sigma}$ are needed, from prior experience or from the plotted data. An initial fit is given by

$$y = \hat{\theta} x.$$

Table 12.2 shows the layout of the calculation. Given the values of the fitted y for this line, denoted by $\tilde{y}_i = \hat{\theta} x_i$, one can derive a set of residuals, $\tilde{e}_i$, from

$$\tilde{e}_i = y_i - \hat{\theta} x_i, i = 1 \text{ to } n.$$

The residuals act as the sample of observations on the distribution defined by the random component of the model, i.e., of $\sigma N(p)$. If we order the residuals, we have to reorder all the data pairs (x_i, y_i) in parallel, and this reordering has to be repeated each time the estimates of the parameters are adjusted. To avoid this we make use of the idea of the rank. The **rank** is the number giving the position of a value in an ordered set of values, from 1 for the smallest to n for the largest.

$n = 20$, $\hat{\theta} = 1.28$; $\hat{\sigma} = 2.51$; $\Sigma |e^*| = 8.33$

			Normal distribution				
y	x	fit	e	rank	p^*	$\hat{y}$	$|e^*|$
10.7	9.5	12.18	−1.48	15	0.28	10.71	0.01
6.5	5.0	6.41	0.09	10	0.52	6.57	0.07
29.4	23.0	29.50	−0.10	11	0.48	29.34	0.06
17.2	15.2	19.49	−2.29	17	0.18	17.21	0.01
18.4	11.4	14.62	3.78	2	0.87	17.43	0.97
19.7	11.8	15.13	4.57	1	0.97	19.71	0.01
16.6	12.1	15.52	1.08	7	0.67	16.64	0.04
29.0	22.0	28.22	0.78	8	0.62	29.00	0.00
40.5	28.2	36.17	4.33	2	0.92	39.65	0.85
14.2	12.1	15.52	−1.32	14	0.33	14.40	0.20

etc.

Table 12.2. Least absolutes fitting of a regression quantile function — crow–car data (normal model)

It can be calculated directly in many statistical programmes and spreadsheets. As a general notation we write that the rank, r, of a value z_i in a set of n values z_1, z_2, ..., z_n is given by a function

$$r_i = \text{RANK}(z_i; z_1, z_2, ..., z_n; n).$$

Thus $r_i = 2$ if z_i is the second smallest value. Some ranking functions give the ranks in descending rather than ascending order and some give either option. Using the rank function, the median-p_i for each residual can be obtained without having to keep reordering the data set. This, as we saw in Section 4.2, is found using $p_i^* = \text{BETAINV}(0.5, r_i, n + 1 − r_i)$. We then have the fitted value for the median of each of the y_i, $\hat{y}_{M,i}$. This can be regarded as the predicted value given by this methodology. It is given for our simple model by

$$\hat{y}_{M,i} = \hat{\theta}x_i + \hat{\sigma}N(p_i^*).$$

It is seen that here the fitted model allows both for the linear regression and for the ordered position of the residual in the distributional element of the model. Having obtained the $\hat{y}_M$ we can calculate the distributional residuals as

$$e_i^* = y_i - \hat{y}_{M,i}.$$

Notice that the distributional residuals are generally smaller than the ordinary residuals since they now take into consideration the shape of the distribution. We can now introduce a third criterion of fit:

$$C_3: \ \Sigma|e_i^*| \ \text{minimum.}$$

This is the distributional least absolutes criterion introduced in Chapter 10.3. Table 12.2 shows the result of choosing both $\hat{\theta}$ and $\hat{\sigma}$ to minimize this. Thus the parameters of both the deterministic and stochastic parts of the model are estimated as part of the same minimization.

The above example gives the simplest of linear regression models. The methodology is the same for the general linear model which can be expressed as

$$Q_y(p\,|\,\underline{x}) = \lambda + \theta_1 x_1 + \theta_2 x_2 + \dots + \theta_k x_k + \eta S(p).$$

Here there are k regressor variables. These should be independent of each other but may be functions, such as x, x^2, x^3, ... Table 12.3 shows the stages of the fitting process.

In Table 12.3 we have simply extended the method of least absolutes to use with regression models. The method of distributional least squares could similarly have been used, with either approximate rankits or the derivation of the exact rankits for the specified distribution $S(p)$.

Three approaches to regression modelling have now been introduced. The differences between them can be seen from how one would go about predicting a quantile for a future value with given x. The ordinary methods of least squares and least absolutes provide for a prediction the future of value of the mean or median of y given a value of x. To find a quantile, further distributional information is needed and a separate analysis and distributional model fitting of the residuals are required. From quantile regression a conditional predictor of, say, the 95th percentile could be obtained, but a separate calculation would be needed for the 90th percentile. For the methods of distributional least squares and absolutes a prediction, for a given x, of any required percentiles can be obtained from the complete fitted model.

Create table	$y, x_1, x_2, \ldots$
Define model	$\lambda + \theta_1 x_1 + \theta_2 x_2 + \ldots + \theta_k x_k + \eta S(p)$
Allocate initial estimates	$\hat{\lambda}, \hat{\theta}_1, \hat{\theta}_2, \ldots, \hat{\theta}_k, \hat{\eta}$
Create fitted values of deterministic component, $\tilde{y}_i$	$\tilde{y}_i = \hat{\lambda} + \hat{\theta}_1 x_{1i} + \hat{\theta}_2 x_{2i} + \ldots + \hat{\theta}_k x_{ki}$
Obtain initial residuals	$\tilde{e}_i = y_i - \tilde{y}_i$
Obtain the ranks of these residuals	$r_i = \mathrm{rank}(\tilde{e}_i; \tilde{e}_1, \tilde{e}_2, \ldots, \tilde{e}_n; n)$
Obtain the median-p_i	$p_i^* = \mathrm{BETAINV}(0.5, r_i, n + 1 - r_i)$
Obtain fitted error terms (median rankits)	$\hat{e}_i = \hat{\eta} S(p_i^*)$
Obtain fitted regression quantiles	$\hat{y}_i = \tilde{y}_i + \hat{e}_i$
Hence distributional residuals	$e_i^* = y_i - \hat{y}_i$
Choose parameters to minimize $\Sigma |e_i^*|$	Obtains estimates and a best $\hat{y}_i$
Plot fit-observation diagram	y_i against $\hat{y}_i$

Note: It can happen that the above minimization leads to confusion between deterministic and stochastic structures. This is shown by the occurrence of a systematic structure in the ranks r_i. In such cases, $\Sigma |\tilde{e}_i|$ and $\Sigma |e_i^*|$ may be minimized separately or in combination.

Table 12.3. Fitting a linear regression quantile model

Thus we might predict the median or predict the interval between, say, the 2.5th and 97.5th percentiles (corresponding to the quantiles of − and +1.96 on the standard normal distribution). The interval is called the **95% prediction interval**. For the data of Table 12.1 consider a "crow flies" line of 5 km. The slope estimate is scale free so we have the predicted median = $1.28 \times 5 = 6.4$ km for the car journey. For the scale parameter we must convert from the unit of a quarter of a kilometre to a km unit. Thus the scale parameter in km is 2.51/4 = 0.627. Hence the 95% prediction interval for the car journey is (6.4 − 1.96 × 0.63, 6.4 + 1.96 × 0.63) = (5.2, 7.6) km.

This section has sought to show the value of using quantile regression models in the analysis of regression data. The example discussed referred to the criteria of least absolutes and least squares. This does not imply that they are the only applicable ones; for example, the method of maximum likelihood could equally well be used to estimate the parameters of a model expressed as a regression quantile function.

Before moving on, it is important to note that the regression quantile formulation puts the deterministic and stochastic elements of a model on the same footing. We may build a model with a construction kit approach applied to both deterministic and stochastic terms. Many statistics texts discuss the building of the deterministic part of the model, e.g., Gilchrist (1984), Walpole and Myers (1989).

Previous chapters have discussed the building of the stochastic element in its quantile function form. Clearly both aspects can now be
seen as parts of the same problem. The model is treated as a single
model for the total behaviour of the data, deterministic and stochastic.

12.2   Quantile autoregression models

A commonly occurring situation is where the distribution of an observation depends on the values taken by previous observations. For
example, the concentration of a batch of a chemical from a batch
production plant will relate most strongly to that of the last batch
produced. If we model these relationships using a linear regression
model, the model is referred to as an **autoregression model**. The
simplest model relating the p-quantile of a new observation to the
observed value of the previous observation is

$$Q(p;x_{t-1}) = \theta x_{t-1} + \eta S(p), \; t = 1, 2, \ldots, t-1, t, \ldots -1 < \theta < 1, \; \eta > 0.$$

Such a model is a **first-order autoregression quantile function**.
The term "first order" refers to the fact that only the last observation
appears on the right-hand side of the expression. A second-order model
would involve $\theta x_{t-1} + \phi x_{t-2}$ in the "regressor" term. If we wanted to
simulate data from this model we would repeatedly use the model on
a sequence of independent random numbers $u_1, u_2, \ldots$, with

$$x_t = \theta x_{t-1} + \eta S(u_t), \; t = 1, 2, \ldots, t-1, t, \ldots, x_0 = 0.$$

The form of this expression is identical to the regression example
of the last section, so it may be fitted and analyzed in exactly the
same manner.

   If $S(p)$ is a symmetric distribution about zero, then the x_t sequence
will have zero mean and will show an autocorrelated wandering behaviour. It can be shown that the autocorrelation coefficient between
observations h apart, the autocorrelation for lag h, is $\rho(h) = \theta^h$. An
interesting feature appears when one allows $S(p)$ to be a heavy-tailed
distribution. When industrial process data is studied over long time
spans such heavy-tailed distributions are often found to occur. The
effect of the occasional large value of $S(p)$ out in the tail is quite
dramatic. The sequence of x values generated by the model will show
a sudden jump. It would appear that there are discontinuities in the

process; something has gone wrong and action had better be taken to put it right. Yet this is natural data from a stable process. The discontinuities are just the effect of having a heavy-tailed distribution in an autoregressive process. Thus, to improve the process, management must work on the causes of the distributional form not on those of the one apparently discrepant observation.

12.3   Semi-linear and non-linear regression quantile functions

The example used to introduce regression quantile models is a linear model; that is to say, it involves the parameters singly in additive terms. Classical statistics uses linear in this context to refer just to the parameters of what is effectively the position parameter of the model. The deterministic part of the model simply controls the position. In the example linear also refers to the scale parameter. We have generalized the term **linear model** slightly to include models that have both the position and the scale controlled by linear sets of parameters. Models must also allow for distributions that have non-linear shape parameters in the regression quantile function. Following the previous terminology these are called **semi-linear models**. Thus we have extended the applicability of the terminology introduced in Section 4.6 to include regression models. The simplest example of the semi-linear model that illustrates the main features takes the form:

$$y_p = \lambda + \theta x + \eta S(p, \alpha).$$

The variable x influences y via the straight line, linear relation $\lambda + \theta x$, and the "error" distribution is given by a quantile function, $\eta S(p, \alpha)$. The parameters are λ, the intercept of the line; θ, the slope; η, the scale of the distribution; and α, the shape parameter of the distribution $S(p, \alpha)$. The merit of this formulation over the standard one is that all the parameters are explicit in the one equation that defines the model. The semi-linear model can be handled in the same way as a linear model. For example, suppose the normal distribution in the crow–car example is replaced by a symmetric lambda distribution, then there will be one additional parameter for which an initial estimate is required and which will be adjusted in the minimization process. Otherwise the procedure is identical to that in Section 12.1.

At its most general a regression quantile function can be written as

$$Q_Y(p) = Q(p;\underline{x}, \underline{\theta}),$$

where the x and θ represent the regressor variable(s) and parameter(s). This general model can be fitted using the methods previously discussed; however, each case needs to be considered carefully. Each structure will raise different problems and procedures for solution. For the rest of this section we will simply illustrate different non-linear models and their study through a number of examples.

Example 12.1: Let us begin by reconsidering the example used to introduce this chapter, which analyzed the Crow–Car data of Table 12.4. The model with a normal error was fitted; Table 12.2 gave the analysis. Unfortunately the model with normal errors is not strictly valid. Suppose we simulated the situation using a variety of x and random numbers, u, then the generating model would be

$$y = \theta x + \sigma N(u).$$

This simulation would occasionally generate values where $y < x$, which is not possible in practice. Here is a situation where it is useful to carry out some conceptual modelling before applying standard methods to the data. Consider therefore what is required of a reasonable model. If we are to use distributional least absolutes as the method of fitting, which we have argued is eminently reasonable, then it would be sensible to have $y = \theta x$ as the line of medians of y given x. Second, we want a distribution that has a clear peak and has the lower threshold of y at $y = x$ to ensure that the condition $y \geq x$ holds. A suitable distributional form for this is the Weibull distribution with shape parameter, β, less than one. We thus have the following relationships:

The model

$$y_p = Q(p|x) = \lambda + \eta[-\ln(1 - p)]^\beta.$$

The threshold $= \lambda = x$, hence

$$y_p = Q(p|x) = x + \eta[-\ln(1 - p)]^\beta.$$

The conditional median is θx, which must have a slope greater than one and so we write θx as $(1 + \phi)x$, $\phi > 0$. Hence

$$x + \eta[-\ln(0.5)]^\beta = (1 + \phi)x,$$

therefore

$$\eta = \phi x/[\ln(2)]^{\beta}.$$

Thus a model that satisfies the conceptual modelling requirements is

$$y_p = x + [\phi x/\{\ln(2)\}^{\beta}][-\ln(1 - p)]^{\beta}.$$

Observe that in this model the regressor variable, x, influences both threshold and scale. Thus although the model is semi-linear in terms of the parameters, the regressor variable appears in two different places. The model will therefore be treated as non-linear and the fitting considered step by step. The model can be rewritten as

$$z = (y/x) - 1 = [\phi/\{\ln(2)\}^{\beta}][-\ln(1 - p)]^{\beta}.$$

This form of the model gives z as an increasing function of p. Such functions we call **ordering functions**. The values of z enable the (x, y) pairs to be ordered, the order of z determining the order of the pair. We can thus use the ranks of the z to obtain the ordered increasing median-p_i^*. Table 12.4 shows the calculations. If initial values are allocated, of say $\hat{\phi} = 0.2$ and $\hat{\beta} = 0.5$, the above formulae enable a fitted $\hat{y}$ to be found and hence the evaluation of the distributional least absolutes criteria of $\Sigma|e^*| = \Sigma|y_i - \hat{y}_i|$. From their starting values the two parameters can be adjusted to give the minimum of this criterion. For the data of the example this minimum is about 20% lower than that for the normal model. Figure 12.3 shows the fit-observation plot of y against $\hat{y}$. Notice that this plot shows the fitted y based on both components of the model, deterministic and stochastic.

Example 12.2: A study involved providing an advisory interview for young men and recording how long they took to respond to the advice. There were 324 men in the study. A provisional analysis was required after the 13th week of the study. At that stage only 74 men had responded. The time taken to respond, y_i, was recorded for these 74. Their ages, x_i, were regarded as a relevant supplementary variable. Past experience suggested that time to respond depended on age, with short times initially with an increase to a fairly stable level for older men. A form for such growth was modelled by the function $1 - \exp\{-\gamma(x - 17)\}$, which approaches one as x increases. The 17 is suggested by the fact that 18 was the youngest age of the study group. The distributional models suggested by previous studies of a similar nature were the exponential and the Weibull. The regression quantile model was thus

$$Q(p;x) = \eta[1 - \exp\{-\gamma(x - 17)\}][-\ln(1 - p)]^{\beta}; \eta, \gamma > 0.$$

Weibull model

$n = 20$	$\hat{\phi} = 0.27$					
	$\hat{\beta} = 0.53$					
	$\Sigma|e^*| = 6.65$					
	fit y $|e^*|$					
y	x	$z = y/x - 1$	rank(z)	p^*	fit y	$|e^*|$
10.7	9.5	0.13	19	0.08	10.35	0.35
6.5	5.0	0.30	10	0.52	6.40	0.10
29.4	23.0	0.28	11	0.48	28.98	0.42
17.2	15.2	0.13	18	0.13	16.96	0.24
18.4	11.4	0.61	2	0.92	17.47	0.93
19.7	11.8	0.67	1	0.97	19.19	0.51
16.6	12.1	0.37	5	0.77	16.97	0.37
29.0	22.0	0.32	9	0.57	28.63	0.37
40.5	28.2	0.44	4	0.82	40.50	0.00
14.2	12.1	0.17	16	0.23	14.04	0.16
11.7	9.8	0.19	12	0.43	12.15	0.45
25.6	19.0	0.35	7	0.67	25.60	0.00
16.3	14.6	0.12	20	0.03	15.40	0.90
9.5	8.3	0.14	17	0.18	9.46	0.04
28.8	21.6	0.33	8	0.62	28.59	0.21
31.2	26.5	0.18	15	0.28	31.30	0.10
6.5	4.8	0.35	6	0.72	6.59	0.09
25.7	21.7	0.18	13	0.38	26.48	0.78
26.5	18.0	0.47	3	0.87	26.59	0.09
33.1	28.0	0.18	14	0.33	33.62	0.52

Table 12.4. Least absolutes fitting of regression quantile function — crow–car data
(Weibull model)

It will be seen that the median is $\eta[1 - \exp\{-\gamma(x - 17)\}][\ln(2)]^\beta$, which for
large x approaches $\eta[\ln(2)]^\beta$. The fitting by distributional least absolutes
is straightforward once the suitable values for p_i^* are allocated to each
pair (x_i, y_i). Given initial estimators of the parameters an ordering variable z_i is defined as

$$z_i = y_i/[1 - \exp\{-\hat{\gamma}(x_i - 17)\}].$$

The z_i are standardized observations that have a straightforward Weibull
distribution. The ranks of the z_i can now be obtained as the basis for
calculating p_i^*. Notice, however, that although the sample corresponds to

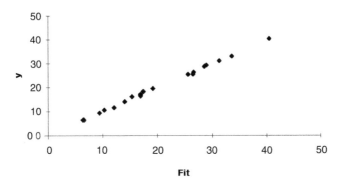

Figure 12.3. Crow–car data — fit observation plot for Weibull regression quantile model

ranks 1 to 74, the value of n is 324. Thus the p^* values for the censored sample available will all be relatively small. The fitting process proceeds as in previous examples. As the exponential is the Weibull with $\beta = 1$, both models can be simply fitted and compared. For the Weibull-based fit the long-term median was 32 with a minimum sum of absolute errors of 15.6, compared with 23.1 when the shape parameter was set at one for the exponential-based model. The high median explains why only a fraction of the times were under the 13-week limit at which the sample was censored.

There is a little practical postscript to this example. The fact that the median of the fitted model is above the value of 13 implies that the main shape and shape changes with age were not clearly observed, as only the very small ordered observations were available. Thus a complex structure was being fitted with only a proportionately small censored sample. This process was like trying to describe a complete picture on the basis of seeing only the corner. Warned by this feature of the data, the model was refitted ignoring the information on the age effect, i.e., with a constant scale parameter. A constant scale corresponds to letting γ become infinite, which was not an option in the optimization previously performed. The modification was thus treated as a fresh model to fit. It was found that using this model as the distribution gave a marginally better fit with one less parameter. Thus although this example has provided a nice illustration of a regression quantile fit, it also illustrates that modelling requires a sceptical and cautious approach to the models that we use.

Example 12.3: In forecasting some variable y, use is often made of a linear trend against time with a normally distributed random variation; however, in some situations this variation may have longer or shorter

tails than a normal, although it will still be symmetrical. The symmetric
lambda distribution provides such a model. Thus using t for time we have

$$Q_Y(p;\ t) = \alpha + \beta t + (\eta/\theta)[p^\theta - (1 - p)^\theta].$$

This model could sensibly be fitted using distributional least absolutes
with discounting. In this case, however, the discounting would be related
to the time in the past at which the observation was obtained (see, for
example, Gilchrist (1976)).

Example 12.4: A paper by Castillo and Hadi (1995) considers the lifetime, y, of materials under a level of stress x, standardized so that $0 \leq x$
≤ 1. A model is justified of the form

$$Q_Y(p;\ x) = (p^\alpha - D - Cx)/(Ax + B).$$

This model is a power distribution with scale and position depending on
stress as the regressor variable.

   The above examples have sought to illustrate some of the forms
that non-linear models may take, and in Example 12.1 the approach
to fitting a model by the method of distributional least absolutes. It is
hoped that the illustrations are enough to show the clarity that is
evident when the regression quantile function shows the whole parametric form of a model. On the basis of the various illustrations it is
clear that the position, scale, skewness, and shape parameters could
all become functions of the regressor variables. The models that we
can fit are sensibly limited to models with reasonable behaviour, which
here manifests itself particularly as the requirement that an ordering
function may be found. The fact that the models are fully parametric
ensures that all the parameters can be estimated by a single minimization of an appropriate criteria.

12.4   Problems

   1.   Some data quoted by Mosteller et al. (1983) gives the
        stopping distance, d, of vehicles as a function of their
        speed, s. It is shown that a good fit is obtained by using
        the square root of stopping distance as the independent
        variable. Suggest a suitable form of model for the case
        where one wishes to keep the stopping distance as the

independent variable. Explore suitable models for the following set of distance–speed data, based on an automated stopping device.

s	10	10	15	15	20	20	25	25	30	30	35	35	40	40	45	45
d	2	2	5	6	13	11	23	21	29	36	49	50	68	71	84	81
s	50	50	55	55	60	60	65	65	70	70	75	75	80	80		
d	107	99	127	132	168	122	211	195	232	176	244	263	269	236		

2. Example 1.20 discussed a model for the pitting of metal due to corrosion; the depth, d, of corrosion is measured at times t. Fit the model given to the following data:

t	1	2	3	4	5	6	7
d	1.05	1.55	1.49	2.25	2.75	2.49	3.36
t	8	9	10	11	12	13	14
d	3.32	2.19	4.14	4.30	4.56	4.83	5.10
t	15	16	17	18	19	20	21
d	5.04	4.92	5.48	4.46	5.53	6.07	6.47
t	22	23	24	25	26	27	28
d	4.54	6.88	6.93	7.15	6.06	6.70	7.59

3. The data below gives some daily humidity, h, and sunshine, s, for a weather station in Sheffield.
    (a) Fit the model $Q(p) = 100 - \eta[-\ln(1-p)]^{\beta}$ to the humidity data using distributional least absolutes.
    (b) Add the regression term θs to the model and refit. Compare the quality of fit of (a) with (b).
    (c) Replace the term θs with the term $\theta \bar{s}$, where $\bar{s}$ is the previous day's sunshine figure. Again compare the fit.
    (d) Replace the term θs by an autoregression term $\theta \bar{h}$, where $\bar{h}$ is the previous day's humidity. Yet again compare the fit.

$h\%$	64	100	99	78	70	99	75	80
s hrs	6.7	0.0	0.0	5.8	4.3	2.3	9.8	1.8
$h\ \%$	72	72	93	65	91	96	95	80
s hrs	9.9	6.5	2.6	9.6	0.0	0.2	3.5	1.5
$h\%$	70	86	71	77	60	63	79	78
s hrs	1.1	2.0	9.8	2.7	15.3	8.1	1.4	3.7
$h\%$	67	88	74	62	71	75		
s hrs	8.5	1.5	6.2	7.4	6.7	2.8		

4. The time to breakdown of insulation can be modelled by
    an exponential distribution with zero threshold and a scale
    parameter η/v^{β}, where v is the voltage across the insula-
    tion. On testing 99% of the test pieces lasted 2000 hours
    at 1000 volts and 1100 hours at 2000 volts. What is the
    maximum voltage allowable if it is required that 90% of
    the items last at least 2500 hours?

CHAPTER 13

Bivariate Quantile Distributions

13.1 Introduction

For a single variable X we can draw the quantile function, $Q(p)$, to show how x relates to p. We have also seen that $Q(p)$ can be viewed as the transformation of a uniform variable to generate X. Suppose we now move to problems with two variables, X and Y, the bivariate situation. If we replace X by two variables (X, Y), we need to replace p by (p, r), a pair of values, each in the range $(0, 1)$. We may then define a bivariate distribution based on quantiles by

$$x = Q_x(p, r)$$
$$y = Q_y(p, r),$$

where $Q_x(p, r)$ and $Q_y(p, r)$ are **bivariate quantile functions**. We thus transform points in the (p, r)-plane, the unit square, to points in the (x, y)-plane, as Figure 13.1 illustrates. As with the univariate case we may imagine the (x, y) data as simulated by pairs of independent uniform values, (p, r), being substituted in the quantile functions. Curves in (p, r) will transform to curves in (x, y). Of particular interest will be the two quantile curves corresponding to $p = p_o$ and $r = r_o$, where p_o and r_o are constant values, such as 0.95 for outer quantiles. The region to the left of $p = p_o$ will transform to a region A_p in the (x, y)-plane such that

$$\text{Prob}[(x, y) \text{ is in } A_p] = p_0,$$

with a similar interpretation for $r = r_o$.

In terms of cumulative distribution functions the univariate function $F(x)$ is replaced by a **bivariate cumulative distribution function**, $F(x, y)$, defined by

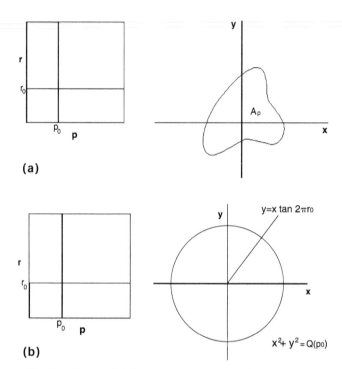

Figure 13.1. The bivariate model form (a) the general model; (b) the circular family

$$F(x, y) = \text{Prob}(X \le x, \ Y \le y).$$

If we set $y = \infty$, this gives simply the cumulative distribution function for x, $F(x)$, called the **marginal CDF of X**. A further important cumulative distribution function is the **conditional CDF of X** given y, $F(x|y)$. This defines the conditional distribution of X for a fixed value of $Y = y$. From the definition of conditional probability (Section 6.8.1), this can be expressed as

$$F(x|y) = [\partial F(x, y)/\partial y]/[\partial F(y)/\partial y].$$

The bivariate probability density function $f(x, y)$ is obtained by differentiating $F(x, y)$ with respect to both x and y.

There are a variety of approaches to relating the (x, y) values to the (p, r) to create bivariate quantile functions. We will illustrate two forms of the model:

The general form: $x = Q_x(p, r)$, $y = Q_y(p, r)$.

The marginal/conditional form: $x = Q_x(p)$, $y = Q_y(p, r)$.

It will be seen that in the marginal/conditional form the first expression is a univariate quantile function for the marginal distribution of x. As a consequence of this, if x is fixed, then p becomes fixed; so the quantile function for y, as a function of r, is the quantile function for the conditional distribution of y given x. This form has an alternative formulation obtained by interchanging x and y and starting with the marginal distribution of y. It may be noted that all the main features of obtaining density functions by differentiation, of substitution to get p-densities in terms of, here, p and r, etc., are analogous between the bivariate and univariate situations. Thus substituting the above quantile functions in a bivariate PDF, $f(x, y)$ will give the (p, r)-PDF, $f_{p, r}(p, r)$. Appendix 3 gives some further detail.

The aim of this chapter is to extend to bivariate distributions some of the quantile function modelling of the previous chapters. We do this by illustrating the structure and fitting of a number of quantile-based bivariate distributions.

13.2   Polar co-ordinate models

The circular distributions

One of the simplest models in general form is based on the use of polar co-ordinates. Here, thinking in simulation terms, p generates radial distances from the origin to the point (x, y) and r generates the rotational angle from the x axis. The basic form of the model can be written as

$$x = [Q(p)]^{1/2}\cos(2\pi r), \qquad 0 \le p, r \le 1$$
$$y = [Q(p)]^{1/2}\sin(2\pi r),$$

where $[Q(p)]^{1/2}$ is a quantile function for a univariate distribution of distributional range $(0,\infty)$ or $(0,$ some positive constant$)$. Figure 13.1(b) shows the model as a transformation from the uniform square for (p, r) to the plane of (x, y). The p-quantile curve is given by looking for the curve in the (x, y)-plane that corresponds to $p = $ constant, i.e., the curve that does not contain the effect of r. Using the Pythagorean-based result that $\sin^2\theta + \cos^2\theta = 1$ gives

$$x^2 + y^2 = Q(p) \text{ all } r.$$

This is thus a circle with probability p of observations lying inside it. The squared radius of the circle is given by the quantile function $Q(p)$. The r-quantile curve is given by removing the influence of p to give

$$y = x \tan(2\pi r) \text{ all } p$$

(see Figure 13.1(b)). Important cases of these quantile curves are the **median curves**. The p-median curve is

$$x^2 + y^2 = Q(0.5) \text{ all } r.$$

It is evident from the definition that half the joint probability lies inside this circle. Similarly, the r-median curve is

$$y = 0, \text{ since } \tan(2\pi/2) = 0.$$

Thus half the probability lies above and half below the x-axis.

It is now evident that probabilities are calculated by looking at regions of the (x, y)-plane and the corresponding regions of the (p, r)-plane. For example, to find the probability of an observation, (x, y), lying in the shaded region of the (X, Y) plane of Figure 13.1(b) we calculate the probability of (p, r) lying in the corresponding shaded region of the (p, r)-plane, which for this jointly uniform distribution is simply $p_0 r_0$. In Appendix 3 the formulae for defining and transforming joint PDF are given and illustrated for these models. It is shown there that the circular model of the (p, r) form of the joint density of (x, y), the p, r-joint PDF, is given by

$$f_{p,r}(p, r) = 1/[\pi q(p)].$$

The fact that this does not involve r, but only p, underlines the circular nature of the distributions defined in this way.

The fact that bivariate models can be regarded as generated from independent pairs of uniforms can be used in several ways. We illustrate two:

(a) The models can be used to explore the form of distributions by simulation. Figure 13.2(a) shows 100 pairs of random numbers. In Figure 13.2(b) these have been substituted in the model being considered to give simulated points in the (X, Y) plane.

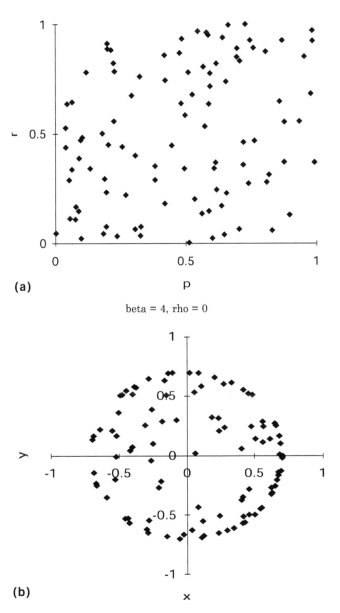

(a)

beta = 4, rho = 0

(b)

Figure 13.2. Simulation of a bivariate distribution

(b) Treating p and r as uniforms enables the expectations to be evaluated simply. For example,

$$E(X) = E[\{Q(p)\}^{1/2}\cos(2\pi r)]$$
$$\quad = E[\{Q(p)\}^{1/2}]E[\cos(2\pi r)], \quad \text{since } r \text{ and } p \text{ are independent.}$$
$$\quad = 0, \quad\quad \text{since by symmetry the second term has zero mean.}$$

We similarly have $E(Y) = 0$. A consequence of these results is that

$$V(X) = E(X^2) = E[Q(p)]\ E[\cos^2(2\pi r)]$$

which leads to

$$V(X) = (1/2)E[Q(p)], \text{ with the identical result for } V(Y).$$

A further expectation involving both (x, y)-plane variables is $E(XY)$. This can be evaluated in exactly the same way, thus

$$E(XY) = E[Q(p)]E[\cos(2\pi r)\sin(2\pi r)].$$

As the term $\cos(2\pi r)\sin(2\pi r)$ is symmetrical about zero for r going from 0 to 1, the last expectation is zero; hence $E(XY) = 0$, thus X and Y are uncorrelated.

The Weibull circular distribution

As the variances of X and Y are both $(1/2)\ E[Q(p)]$ and a natural form for $Q(p)$ would be some form of decaying distribution, a first simple model to use would be to make $Q(p)$ an exponential distribution with scale parameter 2. This would give $Q(p) = 2[-\ln(1 - p)]$, with therefore $V(X) = V(Y) = 1$. The radial distribution is thus a Weibull distribution. The p-quantile curve is found by squaring and summing the two quantile functions to give

$$x^2 + y^2 = Q(p) = -2\ln(1 - p)$$

and hence

$$1 - p = \exp[-(1/2)(x^2 + y^2)].$$

Substituting to get the (p, r) joint density, as derived in Appendix 3, we get

$$f_{p,r}(p, r) = 1/[\pi q(p)] = (1/2\pi)(1 - p)$$

and hence

$$f(x, y) = [1/\sqrt{(2\pi)}] \exp(-(1/2)x^2); [1/\sqrt{(2\pi)}] \exp(-(1/2)y^2).$$

Thus the model is the product of two independent standard normal distributions.

The form of distribution given is a standard form with zero means and unit variances. It is evident from this discussion that the move to a non-standard form of this family of distributions would be obtained by writing

$$x = \lambda_x + \eta_x[Q(p)]^{1/2}\cos(2\pi r), \qquad 0 \le p \le 1, 0 \le r \le 1.$$
$$y = \lambda_y + \eta_y[Q(p)]^{1/2}\sin(2\pi r),$$

using position and scale parameters in an obvious notation. It follows from the zero means and unit variances that for this particular Weibull family the λ are the means and the η are the standard deviations.

We have obtained the bivariate normal from the general circular distribution by using for $Q(p)$ an exponential distribution. The exponential is, as we have seen, a special case of the Weibull distributions. If we take the general case of arbitrary β but put in the necessary constant to keep the unit variances we have

$$Q(p) = [2/\Gamma(\beta + 1)][-\ln(1 - p)]^\beta.$$

The term $\Gamma(\beta + 1)$, which is a gamma function, is the mean of the standard Weibull distribution. This model is a general circular bivariate distribution with zero means and unit variances, but is only bivariate normal for the case $\beta = 1$. We have termed our general model the circular Weibull distribution. By way of illustration, if we set a value for β of less than one, the Weibull Distribution has a modal value at $(1 - \beta)^\beta$. The consequence of this is that the (x, y) points tend to cluster about the modal circle, as in Figure 13.3(a) where $\beta = 0.2$.

The generalized Pareto circular distribution

As a further example we make use of the generalized Pareto distribution for $Q(p)$. Again we need a suitably chosen constant to give the unit variances for X and Y. The form of the distribution is

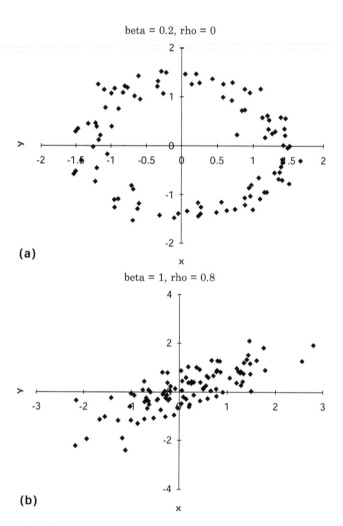

Figure 13.3. The Weibull family: (a) and (b) circular

$$Q(p) = 2[1 - (1 - p)^\beta]/\beta.$$

For negative β this distribution has long tails. Thus there is a central concentration of points in the x-y scatter diagram with a small scatter of points well out. For the distribution to have non-infinite variances for X and Y we must have $\beta > -0.5$, although the distribution still exists for all values of β. For the limiting case of a zero value of β, the shape of $Q(p)$ is the exponential distribution. A consequence of

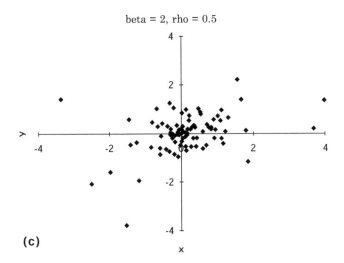

Figure 13.3. The Weibull family: (c) elliptical forms

this is that again the bivariate normal is a special case of the distribution. For positive β the distribution has a maximum of $2/\beta$ so the maximum radius is $\sqrt{(2/\beta)}$. For $\beta = 1$ the distribution is shaped as a disc of radius $\sqrt{2}$. For $\beta > 1$ the generalized Pareto distribution becomes an increasing function up to its maximum radius. The effect of this was, in fact, seen in Figure 13.2(b), which shows a member of this family with $\beta = 4$.

The elliptical family of distributions

A natural generalization of the circular family is given by

$$x = [Q(p)]^{1/2}\cos(2\pi r), \qquad 0 \leq p, r \leq 1$$
$$y = [Q(p)]^{1/2}[\tau\sin(2\pi r) + \rho\cos(2\pi r)],$$

where

$$\tau^2 + \rho^2 = 1.$$

A little algebra and trigonometry lead to

$$x^2 - 2\rho xy + y^2 = (1 - \rho^2)Q(p),$$

so the p-Quantile curve is an ellipse.

The moments follow much as for the circular family with

$$E(X) = E(Y) = 0 \text{ and } V(X) = V(Y) = E(Q(p)]/2.$$

However, here we have

$$E(XY) = \rho E(Q(p)]/2,$$

so the correlation between X and Y is ρ.

Example 13.1: Returning to the exponential form for $Q(p)$, $-2\ln(1-p)$, of the earlier example we obtain the standard bivariate normal with correlated X and Y. The p-quantile curves are given by the ellipses

$$x^2 - 2\rho xy + y^2 = -2(1 - \rho^2)\ln(1 - p).$$

Using the simulation model of the previous section we get the x-y scatter plot shown in Figure 13.3(b) for the elliptical Weibull with $\beta = 1$, i.e., the bivariate normal. In part (c), the $\beta = 2$ leads to a long-tailed form with widely scattered observations.

If the original data distribution is not standard, then it must first be standardized. The sample values of means variances and correlations give reasonable estimates of the corresponding population statistics. These can be used to roughly standardize the two variables to give x and y. The value of

$$z = \sqrt{[\{x^2 - 2\rho xy + y^2\}/(1 - \rho^2)]}$$

can then be calculated and ordered. If the bivariate normal is the correct distribution, then the Z will have the Weibull distribution given by

$$[Q(p)]^{1/2} = [-2\ln(1 - p)]^{1/2}.$$

The above analysis relates to the p-quantile curves. The r-quantiles can also be found. Assuming standardized data and eliminating $Q(p)$ from the original equations of the elliptical model leads to

$$r = (1/2\pi)\tan^{-1}[\{y/x - \rho\}/\{\sqrt{(1 - \rho^2)}\}] \; x \geq 0.$$

The fact that $\tan^{-1}(.)$ is not a single-valued function means that some care is needed to choose the r-values that are the correct radii. The ordered values of r can be plotted directly against the p-rankits, for a median-based plot, or against the expected values for the order statistics of the uniform distribution, $i/(n + 1)$.

The discussion thus far has used simply the exponential $Q(p)$. If we use the Weibull or the generalized Pareto forms, we obtain correlated forms of the models already discussed in the previous section.

13.3   Additive models

We illustrate additive models and also some of the methods of studying bivariate quantile distributions by considering the errors in measurement model. Suppose we have a variable X with some distribution $Q(p)$. We can only observe X using equipment with an inherent error, E, whose distribution is $S(r)$, independent of X. The observations, Y, are thus modelled by

$$y = x + e.$$

We can thus write

$$x = Q(p), \ e = S(r) \text{ and hence } y = Q(p) + S(r),$$

where p and r are independently and uniformly distributed. Interest here focuses on the transformation from the (p, r)-plane to the (x, e)- and (x, y)-planes. For the first of these it is evident that the lines $p = p_0$ and $r = r_0$ transform directly to lines $x = Q(p_0)$ and $e = S(r_0)$. The conditional distributions of y given e_0 and y given x_0 are

$$y_p|\, e_0 = Q(p) + S(r_0)$$

and

$$y_r|\, x_0 = Q(p_0) + S(r).$$

If we consider the (y, x)-plane, which is the one of prime practical interest, then we need consider probabilities such as $\text{Prob}(y \leq y_0)$. The line $y = y_0$ corresponds to a curve, C_0, in the (p, r)-plane. In theoretical terms we have

$$Q(p) = y_0 - S(r)$$

so

$$p = Q^{-1}[y_0 - S(r)].$$

This is the p, r relation that defines C_0. The area enclosed by C_0 gives the probability so

$$\text{Prob}(y \le y_0) = \int_0^1 Q^{-1}[y_0 - S(r)]dr.$$

A numerical approach to this is to fill the (p, r)-plane with a uniform lattice of points. For each point (p_i, r_j) the value $y_{(i, j)}$ is obtained and the proportion of $y_{(i, j)}$ less than y_0 used to approximate $\text{Prob}(y \le y_0)$. By using this device for each r_i the curve C_0 can also be approximated.

We often need the reverse information, i.e., given knowledge of y, what can be said about x. For example, we might be interested in $\text{Prob}(x \le k | y \le k)$ where k is a constant. Again, this can be approximated using the counting approach described above.

13.4   Marginal/conditional models

Two marginal distributions on their own do not uniquely define a bivariate distribution; however, there are a number of models that use a marginal and a conditional pair in quantile form to define a bivariate distribution.

Example 13.2: The bivariate logistic distribution is defined by the pair of quantile functions

$$Q_x(p) = \lambda_x + \eta_x \ln[p/(1-p)],$$
$$Q_{y|x}(r|p) = \lambda_y + \eta_y \ln[p\sqrt{r}/(1-\sqrt{r})].$$

It will be seen that the marginal distribution of x is a logistic, but the conditional distribution of y given x is a p-transform of the logistic (see Castillo, Sarabia and Hadi (1997)). The link of the model to regression is seen by noting that the median of the conditional distribution for given p is

$$M = \lambda_y - \eta_y \ln(\sqrt{2} - 1) + \eta_y \ln p.$$

Notice that the quantiles of y given x, i.e., given p, $y_{|p}$, can be expressed in the relation

$$z = y_{|p} - \lambda_y - \eta_y\ln(p) = \eta_y\ln[\sqrt{r}/(1 - \sqrt{r})].$$

The right-hand side here is an increasing function of r and so provides the ordering function, z. This is used as the basis for deriving the rank order needed for the median rankits for r.

Clearly we could define the model in the reverse fashion in terms of the marginal distribution of y and the conditional distribution of x given y.

Example 13.3: If a centred Pareto distribution is developed, using $q = 1 - p$ and $s = 1 - r$ for simplicity, it takes the form

$$Q_x(q) = \eta_x(1/q^\beta - 1)$$

$$Q_{y|x}(s|q) = (\eta_y/q^\beta)(1/s^{\beta^*} - 1), \qquad \text{where } \beta^* = \beta/(1 + \beta).$$

Here both marginal and conditional distributions are centred Pareto, but with differing scale and shape parameters.

13.5   Estimation

There are a variety of approaches to estimating the parameters of bivariate distributions, paralleling the methods for univariate. We will keep to the method of least absolutes as it is fairly robust, straightforward to implement and convenient in terms of models defined by quantiles. Consider a set of data (x_k, y_k), $k = 1$ to n. The models with initial parameter values can be used to fit values for both x_k and y_k, giving the equivalents of the median rankits denoted by $(M_{x,k}, M_{y,k})$. The least absolutes criterion now becomes

$$C = \Sigma[|x_k - M_{x,k}| + |y_k - M_{y,k}|].$$

As with univariate estimation we evaluate this quantity for the initial parameter values and then adjust them to minimize C. The complication arises, as it did in the fitting of regression lines, that the ranks of the x_k and y_k are different. There is therefore a need for procedures to carry out the ranking. These vary with the form of model. For the polar co-ordinate form of the model the x-y co-ordinate data is transformed to polar co-ordinates and then can be separately ranked. Table 13.1 shows the sequence of operation for the circular distributions.

General Model	$Q_x(p, r) = \lambda_x + \eta_x\sqrt{Q(p;\beta)}\cos(2\pi r)$
	$Q_y(p, r) = \lambda_y + \eta_y\sqrt{Q(p;\beta)}\sin(2\pi r)$
Set initial parameter values	$\lambda_x, \eta_x, \beta, \lambda_y, \eta_y$
Standardize data using these	$x'_k = (x_k - \lambda_x)/\eta_x,\ y'_k = (y_k - \lambda_y)/\eta_y$
Change to polar co-ordinates (z, t)	$z_k^2 = x'^2_k + y'^2_k,\ k = 1, ..., n$
	$t_k = (1/2\pi)\tan^{-1}(y'_k/x'_k),\qquad x'_k > 0$
	$\quad = 0.5 + (1/2\pi)\tan^{-1}(y'_k/x'_k),\ x'_k < 0$
Find ranks of z_k and t_k	$i = \text{Rank}(z_k^2;\ z_1^2, z_2^2, ..., z_n^2;n)$
	$j = \text{Rank}(t_k;\ t_1, t_2, ..., t_n;n)$
Find median-p and median-r	$p_k = \text{BETAINV}(0.5, i, n + 1 - i)$
	$r_k = \text{BETAINV}(0.5, j, n + 1 - j)$
Calculate median rankits	$M_{x,\,k} = \lambda_x + \eta_x\sqrt{Q(p_k;\beta)}\cos(2\pi r_k)$
for x_k and y_k	$M_{y,\,k} = \lambda_y + \eta_y\sqrt{Q(p_k;\beta)}\sin(2\pi r_k)$
Evaluate criterion, choose	$C = \Sigma[|x_k - M_{x,\,k}| + |y_k - M_{y,\,k}|]$
parameters to minimize C	
Plot fit-observation diagrams	$(M_{x,\,k}, x_k), (M_{y,\,k}, y_k)$

Table 13.1. Algorithm for fitting a circular model

For the elliptical form of model there is an additional correlation parameter to initially set. Once the data is standardized the set correlation is used to define the ordering function

$$z^2 = x'^2 - 2\rho x'y' + y'^2$$

in place of the z^2 in Table 13.1. This is then used to rank the z and find the median-p. Similarly, the ordering function t, adjusting for the many valued nature of $\tan^{-1}(.)$, is

$$t = (1/2\pi)\tan^{-1}[\{y'/x' - \rho\}/\{\sqrt{(1-\rho^2)}\}]\qquad x \geq 0$$
$$= (1/2\pi)\tan^{-1}[\{y'/x' - \rho\}/\{\sqrt{(1-\rho^2)}\}] + 0.5,\quad x < 0.$$

This is used to rank the data and find the median-r. The process of Table 13.1 is otherwise unaltered.

For the marginal/conditional form of model, such as the bivariate logistic of Example 13.2, the procedure of Table 13.1 has to be modified. As there are two forms of models, one with x as given the marginal distribution and the other with y, one has to be chosen initially. There

Model for marginal x	$Q_x(p) = \lambda_x + \eta_x \ln(p/(1-p))$
Model for conditional $y \mid x$	$Q_{y \mid x}(r \mid p) = \lambda_y + \eta_y \ln[p\sqrt{r}/(1-\sqrt{r})]$
Set initial parameter values	$\lambda_x, \eta_x, \lambda_y, \eta_y$
Find ranks for x data	$i = \mathrm{Rank}(x_k; x_1, x_2, ..., x_n; n)$, $k = 1, ..., n$
Hence median-p and median rankit	$p_k = \mathrm{BETAINV}(0.5, i, n+1-i)$, $M_{x,k} = Q_x(p_k)$
Create ordering function, z_k for $y \mid x$	$z_k = y_k - \lambda_x - \eta_x \ln(p_k)$
Find ranks of z_k	$j = \mathrm{Rank}(z_k; z_1, z_2, ..., z_n; n)$
Find the median-r	$r_k = \mathrm{BETAINV}(0.5, j, n+1-j)$
Hence median rankits for y_k	$M_{y \mid x, k} = \lambda_y + \eta_y \ln[p_k\sqrt{r_k}/(1-\sqrt{r_k})]$
Evaluate criterion, choose parameters to minimize C	$C = \Sigma[\,\lvert x_k - M_{x,k}\rvert + \lvert y_k - M_{y \mid x,k}\rvert\,]$
Plot fit-observation diagrams	$(M_{x,k}, x_k)$, $(M_{y,k}, y_k)$

Table 13.2. Algorithm for fitting a marginal-conditional model — bivariate logistic

is also the need to use the idea of the ordering function to find the median rankit for the conditional distribution. Table 13.2 shows the procedure for the fitting of the bivariate logistic. Table 13.3 shows part of a tabular layout illustrating the practical sequencing of the procedure. Initial values have to be set for the four parameters and the procedure of Table 13.2 is worked through in the columns of the table. Thus the fitted marginal x is obtained in column 5, the ordering function is used in columns 6 to 8, leading to the fitted conditional values of y in column 9. Now given the data and fitted values, the two sets of distributional residuals, denoted e_x and e_y, are found and their absolute values summed to give the final criterion value. The parameters are then altered to search for the minimum criterion value, which is the quoted value.

If there is no specific reason in the application to choose the x marginal or y marginal at the start of the method, the two may be combined (see Castillo, Sarabia and Hadi, 1997). Table 13.4 gives the steps of the method, which carries out the steps of Table 13.2 twice to obtain two sets of median rankits. These are then treated as predictors of the observed values and a joint predictor is created as a weighted sum. Figure 13.4 shows the fit-observation plots for some data fitted by this method.

$n = 16$; $\lambda_x = 3.07$; $\lambda_y = 0.80$; crit = 0.953; $\eta_x = 0.16$; $\eta_y = 0.03$; $\Sigma|e| = 0.821, 0.132$.

1	2	3	4	5	6	7	8	9	10	11
			x fit			y|x fit				
k	x	y	p_k	$M_{x,k}$	z_k	rank z	r_k	$M_{y\|x,k}$	$|e_{x,k}|$	$|e_{y\|x,k}|$
1	2.64	0.70	0.042	2.57	0.29	6	0.35	0.71	0.075	0.082
2	2.72	0.81	0.103	2.72	0.84	14	0.84	0.81	0.000	0.000
3	2.78	0.75	0.164	2.81	0.29	5	0.29	0.75	0.032	0.032
4	2.84	0.76	0.225	2.87	0.30	7	0.41	0.77	0.029	0.042
5	2.85	0.74	0.286	2.92	0.14	2	0.10	0.74	0.072	0.080
6	2.97	0.80	0.347	2.97	0.54	10	0.59	0.81	0.000	0.007

etc.

Table 13.3. Layout for fitting a bivariate logistic

Work through method of Table 13.2	Obtain $M_{x,k}$ and $M_{y\|x,k}$
Interchange x and y and repeat [ensuring that the k refer to the same pair (x_k, y_k)]	Obtain $M_{y,k}$ and $M_{x\|y,k}$
Combine with weighting γ	$M_{cx,k} = \gamma M_{x,k} + (1-\gamma)M_{x\|y,k}$ $M_{cy,k} = \gamma M_{y,k} + (1-\gamma)M_{y\|x,k}$
Evaluate criterion, choose parameters and weighting to minimize C	$C = \Sigma[|x_k - M_{cx,k}| + |y_k - M_{cy,k}|]$

Table 13.4. Additional algorithm for a combined method for marginal/conditional models

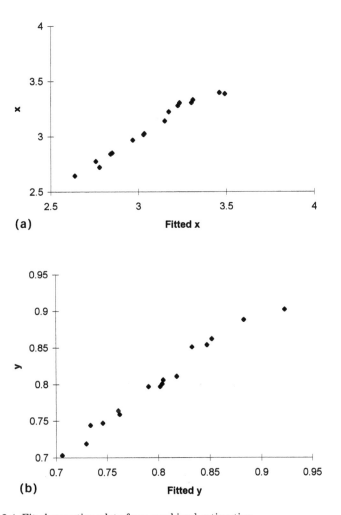

Figure 13.4. Fit-observation plots from combined estimation

13.6 Problems

1. Show that the generalized Pareto marginal-conditional model of Example 13.3 has a marginal distribution for y of the generalized Pareto form.

2. Simulate two sets of 100 uniforms and use them to explore the shapes of the distribution given in Example 13.3. Esti-

mate the parameters of the model from a set of simulated data of known parameter values. Explore the effect of changing β on the criterion of fit.

3. Fit the marginal/conditional form of the bivariate logistic distribution to the following data using least absolutes. Fit with x as the marginal and then y. Fit by the combined method.

x	30.5	24.4	13.6	20.0	31.8	15.3	29.0
y	11.1	8.6	11.3	7.7	10.8	9.0	10.1
x	18.8	10.6	27.5	23.5	24.4	26.4	30.6
y	8.4	12.1	12.4	10.3	9.9	10.3	9.7
x	20.9	17.9	21.6	15.8	18.4	22.0	27.7
y	11.0	8.9	9.4	12.4	9.1	8.9	11.5
x	3.5	35.3	24.8	26.4	5.6	17.7	24.5
y	7.0	10.3	10.4	9.7	9.3	12.0	11.4

4. A small bird migration survey, made from a hide, records for each bird seen the distance, z, and the angle from the north, r. An (x, y) plot of sightings is then constructed. A circular model would be appropriate if it did not ignore the direction of migration. To allow for this, a p-transformation is suggested for the r variable in the circular model. It is proposed to replace $2\pi r$ in the sin( ) and cos( ) terms by $H(r) = \theta + \pi[r^{\alpha} - (1 - r)^{\alpha}]$. Explore this model and suggest improvements. Simulate the model using the exponential for the distribution of z.

CHAPTER 14

A Postscript

The flow chart of Figure 14.1 shows a slightly extended version of the structure of statistical modelling previously described. It shows the main links but not all the possible means of iteration. To round off our study it is worth just looking at each of the elements in the modelling process to see what an approach using quantiles contributes to the classical modelling stages.

The data environment. This term emphasizes that we never have a set of data that does not have a background. It must have been collected in some fashion for some purpose. Something must be known about the variables being measured. The situation is probably not totally unique, someone will have collected data of a roughly similar nature before. There may even be papers or books on the type of problem being studied. All this information can contribute to our understanding of how to approach the details of the modelling exercise, although they should not prevent us from noticing the truly new and unique features of the data.

The data. We have limited ourselves almost entirely to continuous variable data. Quantile function methodology can be applied to discrete variables, but it requires more complex definitions and leads to quantile functions of step form that are not so simple to use. Note that, following Galton's perception of 100 years ago, the ordered data has proved to be central to the methods of data analysis.

Data properties. Classical statistics emphasizes moments as data summaries. These do not always exist. We have shown that there is an extensive range of quantile-based measures that describes well the shapes of samples and populations and which always exists. In particular we have emphasized the measurement of skewness and of shape, with sample functions

A Modelling Process

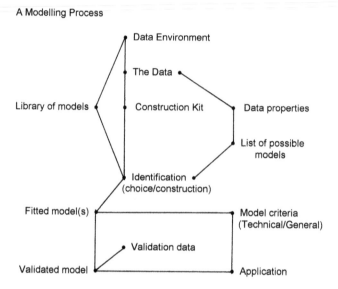

Figure 14.1. The modelling process

such as $t(p)$ and $g(p)$, and also the measures that look at the tails separately.

Construction kit. The prime contribution of quantile functions is their ability to act as the basis for a construction kit for distributions. We have looked at various ways in which the construction kit can be used. We now have rules for construction involving addition, multiplication and transformation, both Q- and p-transformation. Using addition we have shown the value of the reflection family of distributions. These have underlined the value of having five meaningful parameters in a distribution and put the emphasis on skewness and the two tail shapes rather than on skewness and kurtosis. Multiplication has produced a number of distributions of some flexibility. Transformation, both Q and p, in relation to distributions and regression quantile models, has shown the simplicity and power of this way of developing models. It has also emphasized that we do not need to transform data; we keep the natural data and transform the quantile function. As with all construction kits the best way to learn to use them effectively is to play with them frequently.

The library of models. Not all distributions can be expressed explicitly as PDF or CDF. We have seen a number of distributions that add to the library of useful models and which are defined simply by their quantile functions. Our playing, by building simple structures and models and finding out how they behave, has begun to teach us the properties of different models. We have illustrated this with examples in the text and the chapter problems. It has, hopefully, been observed that some aspects of statistics are more elegantly expressed in quantile function form, e.g., distributions of largest and smallest observations, truncated distributions and median rankits. These features also help build the library of structures and models. Two important sections of the developing library discussed in previous pages are the regression quantile models, with their unification of deterministic and stochastic elements, and the basic bivariate quantile models.

List of models. Using empirical evidence of data, the background knowledge of the data environment that may lead to conceptual models, and suitable models from the library, a list of possible models for the specific data begins to form. This now includes quantile functions as well as CDF and PDF. It has also become apparent that because there are so many models, the modeller should be very cautious about moving too quickly to any "one true model." It is much better to hold a small number of tentative models.

Model criteria (technical/general). The data environment and the nature of the data itself must provide the criteria to be used in deciding what constitutes a good model. To these we add statistical criteria and seek some general approach that applies to the complete model and not just to the deterministic or stochastic part of it. Quantile models lead naturally to this comprehensive approach. In the practical examples we have mainly used the method of distributional least absolutes, because it is rather neglected in the textbook literature, is more robust than many other methods, and illustrates well the use of quantile functions.

Identification. Chapter 8 showed the very wide range of plotting techniques based on the ordered data that are available to investigate the various features of distributional shape. These are used to narrow the list of possible distributions or to suggest component shapes that might be used to build a model. The process of identification may then be reduced to choosing which in the list of possible models are the closest to the data in some sense. Experience suggests that identification should lead to models and not "The Model." The statistics may give two models that are almost equally good. The data environment may lead to the final choice. The process of identification makes good use of quantile functions via quantile and other plots. The plots used often lend themselves to the study of models with non-linear parameters using the graphics in a dynamic fashion. We have also emphasized the use in plotting of exact median rankits, rather than approximate, mean rankits. The median-based fit leads to clear interpretations of plotted data. It can readily be seen whether or not there is 50:50 scatter of the data about a line of medians. The quantile approach has also led to the possibility of sequential model building. This has the potential to cover both the deterministic and stochastic elements of models.

Fitting. All methods of estimation can be applied with distributions in quantile function form. However, we have seen that the method of percentiles, probability-weighted moments, and of distributional least absolutes or squares give straightforward methods of particular appropriateness to models given by quantile functions. Maximum likelihood can be applied to models with only quantile function form, although with marginally more difficulty. It does, however, lead to the opportunity to give likelihood limits to the estimated parameters. It was noted that the same form of logic can be used to set limits using the alternative criteria of least squares and least absolutes. It has been shown that all the least and maximum methods lend themselves to practical implementation using now common optimizing routines.

Validation. Again, quantile-based plots and approaches provide an important element of validation and are facilitated by expressing the models in quantile function form. Fit-observation plots with limits, the density probability plot, and the exponential spacing control chart all give simple tools for visual validation. The likelihood approach provides for the more formal testing of specified parameters. The tests also relate easily to the approach adopted to laying out maximum likelihood estimation problems and the use of numerical optimizers.

Application. It is observed that the growth of the study of distributions expressed in quantile form has often taken place in the context and literature of specific application areas. This has particularly been the case where more flexible models were needed. We have shown several examples of situations, usually dominated by the assumption of normality, that can be simply generalized by quantile function notation.

Conclusion. There are vast areas of statistical modelling that are well analyzed using PDF and CDF. Indeed there are many areas where quantile function approaches are entirely inappropriate. It is hoped however that there is enough in this introductory text to suggest that seeing statistical modelling from a quantile function perspective can contribute to problem solving with models. This contribution can be by the development of models using the quantile function construction kit. Alternatively, it may just be that seeing problems from a different perspective can generate new ideas.

Some Useful Mathematical Results

Definitions

The Gamma Function

$$\Gamma(z) = \int_0^\infty e^{-x} x^{z-1} dx, \ \Gamma(z+1) = z\Gamma(z), \ \Gamma(1/2) = \sqrt{\pi}$$

and for integer n,

$$\Gamma(n+1) = n! = n(n-1)(n-2)...3.2.1. \ \Gamma(1) = 1.$$

The Beta Function

$$B(\alpha, \beta) = \int_0^1 x^{\alpha-1}(1-x)^{\beta-1} dx = \Gamma(\alpha)\Gamma(\beta)/\Gamma(\alpha+\beta) = B(\beta, \alpha) \ \alpha, \beta > 0.$$

Special cases:
$B(1, \beta) = 1/\beta.$
$B(\alpha, 1) = 1/\alpha.$
$B(2, \beta) = 1/[\beta(\beta+1)], \ B(\alpha+1,\beta) = [\alpha/(\alpha+\beta)]B(\alpha,\beta).$

Incomplete Beta Function

$$B_z(\alpha, \beta) = \int_0^z x^{\alpha-1}(1-x)^{\beta-1} dx. \ 0 \le z \le 1.$$

Often for statistical purposes the ratio I is used where $I(p, \alpha, \beta) = B_p(\alpha, \beta)/B_1(\alpha, \beta)$. This is a CDF. Its inverse is BETA-INV(p, α, β).

Binomial Coefficient $\binom{n}{r} = n(n - 1)(n - 2)\ \ldots\ (n - r + 1)/r! = n!/r!(n - r)!$ integer n and r, $r \leq n$.

Series

$$1/(1 - p) = 1 + p + p^2 + \ldots.$$
$$(1 - p)^m = 1 - mp + m(m - 1)p^2/2! + \ldots.$$
$$e^p = 1 + p + p^2/2! + p^3/3! + \ldots.$$
$$\ln(1 + p) = p - p^2/2! + p^3/3! - \ldots.$$

Definite Integrals

$H(p)$	$\int_0^1 H(p)dp$
$(\ln p)^m$	$(-1)^m m!$ integer m
$(-\ln p)^m$	$\Gamma(m + 1)$, $m!$ for integer m
$p \ln p$	$-1/4$
$p^m(-\ln p)^n$	$\Gamma(n + 1)/(m + 1)^{n+1}$
$p^m(\ln p)^n$	$(-1)^n\ \Gamma(n + 1)/(m + 1)^{n+1}$
$(\ln p)/(1 - p)$	$-\pi^2/6$
$(\ln p)/(1 + p)$	$-\pi^2/12$
$(\ln p)/(1 - p^2)$	$-\pi^2/8$
$\ln[(1 + p)/(1 - p)]$	$-\pi^2/4$
$p \ln(1 - p)$	$-3/4$
$\ln(-\ln p)$	$-E = -0.57722$ Euler's constant
$\ln(1 + p)/p$	$-\pi^2/12$
$\ln p \ln(1 - p)$	$2 - \pi^2/6$

Indefinite Integrals

$H(p)$	$\int H(p)dp$
$\ln p$	$p \ln p - p$
$(\ln p)^2$	$p(\ln p)^2 - 2p \ln p + 2p$
$p \ln p$	$(p^2/2)\ln p - p^2/4$
$p^2 \ln p$	$(p^3/3)\ln p - p^3/9$
$(\ln p)^m/p$	$(\ln p)^{m+1}/(m + 1)$
$1/(p \ln p)$	$\ln(\ln p)$

Further Studies in the Method
of Maximum Likelihood

The theory of the method of maximum likelihood is based upon the use of the probability density function. The likelihood explicitly involves the data through the PDF and this is therefore the natural form to use. However, for a distribution where a quantile function exists, but a PDF does not, we are forced to use the quantile and related functions to study the problem. The objective of this appendix is to show in more detail how this may be done. In Section 9.4 the emphasis was on numerical solutions. Here we will look at some more theoretical considerations, although most practical situations will lead ultimately to numerical and approximate outcomes. A particular need is to obtain formulae for the variances of the maximum likelihood estimators, so that we can have some sense of the precision in estimation.

For illustration we will concentrate on the linear model

$$Q(p) = \lambda + \eta S(p), \text{ where } S(p) \text{ is in } (-\infty, \infty). \tag{A2.1}$$

Here λ is not a threshold parameter. From this we have

$$q(p) = \eta s(p) \text{ and } f_Q(p) = 1/(\eta s(p)) = (1/\eta) f_S(p), \tag{A2.2}$$

where $f_S(p)$ is the p-PDF for $S(p)$. We also have the "observed" $p_{(r)}$, i.e., the p values that would generate the ordered observations from the model, given implicitly by

$$x_{(r)} = \lambda + \eta S(p_{(r)}). \tag{A2.3}$$

The final basic equation is the log likelihood itself

$$\ell = -\Sigma \ln q(p_{(r)}) = -n \ln \eta + \Sigma \ln f_S(p_{(r)}). \tag{A2.4}$$

Notice that

(a) If $S(p)$ is an explicit function of p, then $f_S(p)$ is also an explicit function even if $f(x)$ does not explicitly exist.

(b) $p_{(r)}$ is itself a function of the parameters, as given in (A2.3). Thus ℓ contains parameters both explicitly and implicitly through $p_{(r)}$. In differentiating ℓ we therefore need the derivatives of $p_{(r)}$ with respect to the parameters. These are obtained by differentiating (A2.3) with respect to the parameters, noting that $x_{(r)}$ is a constant, thus

$$\partial(3)/\partial\lambda \quad 0 = 1 + \eta s(p_{(r)})(\partial p_{(r)}/\partial\lambda)$$

hence

$$\partial p_{(r)}/\partial\lambda = -1/\{\eta s(p_{(r)})\} = -(1/\eta)f_S(p_{(r)}), \qquad (A2.5)$$

$$\partial(3)/\partial\eta \quad 0 = 1.S(p_{(r)}) + \eta s(p_{(r)})(\partial p_{(r)}/\partial\eta)$$

hence

$$\partial p_{(r)}/\partial\eta = -(1/\eta)S(p_{(r)})/s(p_{(r)}) = -(1/\eta)S(p_{(r)})f_S(p_{(r)}). \qquad (A2.6)$$

We can now differentiate the log likelihood to obtain

$$\partial\ell/\partial\lambda = \Sigma\{1/f_S(p_{(r)})\}f'_S(p_{(r)})\{-(1/\eta)f_S(p_{(r)})\}$$
$$= -(1/\eta)\Sigma f'_S(p_{(r)}), \qquad (A2.7)$$

where $f'_S(p_{(r)})$ is the derivative with respect to $p_{(r)}$ of $f_S(p_{(r)})$.

$$\partial\ell/\partial\eta = \Sigma\{1/f_S(p_{(r)})\}f'_S(p_{(r)})\{-(1/\eta)S(p_{(r)})f_S(p_{(r)})\} - n/\eta$$
$$= -(1/\eta)[\Sigma S(p_{(r)})f'_S(p_{(r)}) + n]. \qquad (A2.8)$$

Equating these to zero for maximum log-likelihood values and putting estimated values in equation (3) to give estimated $\hat{p}_{(r)}$ gives

$$x_{(r)} = \hat{\lambda} + \hat{\eta}S(\hat{p}_{(r)}), \qquad (A2.9)$$

$$\Sigma f'_S(\hat{p}_{(r)}) = 0, \qquad (A2.10)$$

$$\Sigma S(\hat{p}_{(r)}) f'_{s}(\hat{p}_{(r)}) + n = 0. \qquad \text{(A2.11)}$$

Solving these equations numerically gives the maximum likelihood estimators. In practical terms the solution seeking the maximum, as described in Chapter 9, is a simpler approach. The value of working through the above lies more in the next step. A result in the theory of maximum likelihood estimation is that for large samples of well-behaved distributions the estimators are approximately normally distributed with expectations equal to the true parameter values and with variances given by the forms 1/I, where the information I is defined for some parameter θ by

$$I(\theta) = E[-\partial^2 \ell / \partial \theta^2] \qquad \text{(A2.12)}$$

The expectation is with respect to the random variables X. In our form we have

$$I(\theta) = E[\partial^2 \{\Sigma \ln q(p_{(r)})\} / \partial \theta^2]. \qquad \text{(A2.13)}$$

This expression can be simplified using three features of the situation:

(a) The linearity of the differentiation and expectation operators enables the order to be changed and terms to be differentiated separately.
(b) Although the $p_{(r)}$ relates to the ordered data, the summation is over all values of r, so the x_r can be treated as a set of n independent observations on the distribution.
(c) The random variable X can be treated as generated by $Q(U)$ from the uniform variables U, corresponding to the p in the quantile function.

Using these three features the expression simplifies to

$$I(\theta) = nE[\partial^2 \{\ln q(U)\} / \partial \theta^2], \qquad \text{(A2.14)}$$

where the expectation is now for the uniform distribution. For the present case this gives, using a simplified notation

$$I(\lambda) = nE[\partial^2 \{-\ln \eta + \ln f\} / \partial \lambda^2]$$

which from (5)

$$= nE[\partial\{(1/f)f'(-f/\eta)\}/\partial\lambda]$$
$$= -nE[\partial\{-f'/\eta\}/\partial\lambda]$$
$$= -(n/\eta)E[f''(-f/\eta)]$$

and hence

$$I(\lambda) = (n/\eta^2)E[f''s(U)fs(U)]. \qquad (A2.15)$$

For the scale parameter we have

$$I(\eta) = nE[\partial^2(-\ln \eta + \ln f)/\partial\eta^2],$$

which using relation (6) and simplifying gives first

$$I(\eta) = -nE[\partial\{(1/\eta)\{1 + Sf')/\partial\eta]$$

and then

$$I(\eta) = (n/\eta^2)E[1 + Sf' + sSf'f + S^2f''f].$$

The expectations are now evaluated for any given $S(p)$ using the uniform distribution.

Bivariate Transformations

In introducing the idea that the quantile function, $Q(p)$, transforms a uniform variable to the defined distribution, we made use of the transformation rule that $f(x)dx = h(p)dp = 1.dp$, using the fact that $Q(p)$ is an increasing function and denoting the PDFs as $f(x)$ and $h(p)$. Thus $f(x) = h(p)(dp/dx) = dp/dx$ is the general rule for single variable distributions. A general analogous result for bivariate transformations is that if (p, r) transforms to (x, y) then the joint PDF satisfy the relation

$$f(x, y) = h(p, r)| J(p, r \mid x, y)|$$

where $J(p, r \mid x, y)$ is the Jacobian defined by

$$J(p, r \mid x, y) = (\partial p/\partial x)(\partial r/\partial y) - (\partial r/\partial x)(\partial p/\partial y).$$

Use may also be made of the fact that

$$J(p, r \mid x, y) = 1/J(x, y/p, r).$$

For the approach taken to bivariate distributions $h(p, r) = 1$ for (p, r) lying in the unit rectangle. We thus have the joint distribution of x and y given by

$$f(x, y) = 1/| (\partial x/\partial p)(\partial y/\partial r) - (\partial x/\partial r)(\partial y/\partial p)| .$$

Example

For the model

$$x = \sqrt{Q(p)}\cos(2\pi r)$$
$$y = \sqrt{Q(p)}\sin(2\pi r)$$

$$J(p, r|x, y) = 1/[\{q(p)/(2\sqrt{Q(p)})\}\cos(2\pi r)\sqrt{Q(p)}\cos(2\pi r).2\pi$$
$$+\{q(p)/(2\sqrt{Q(p)})\}\sin(2\pi r)\sqrt{Q(p)}\sin(2\pi r).2\pi]$$
$$= \pi q(p).$$

As $h(p, r) = 1$ for the bivariate uniform distribution, it follows that the bivariate (p, r)-probability density function can be written as

$$f_{p,\,r}(p, r) = 1/[\pi q(p)].$$

References

Abouammoh, A.M. and Ozturk, A. (1987) On the fitting of the generalised lambda distribution to climatological data. *Pakistan J. Statist.*, **3**, 39–50.

Ali, M.M. and Chan, L. K. (1964) On Gupta's estimates of the parameters of the normal distribution. *Biometrika*, **51**, 498–501.

Andrews, D.F. and Herzberg, A.M. (1985) *Data: A Collection of Problems from Many Fields for the Student and Research Worker*, Springer-Verlag, New York.

Arnold, B. C., Balakrishnan, N., and Nagaraja, H. N. (1992) *A First Course in Order Statistics*, John Wiley and Sons, New York.

Balakrishnan, N. (Ed.) (1992) *Handbook of the Logistic Distribution*, Marcel Dekker, New York.

Balakrishnan, N. and Cohen, A.C. (1991) *Order Statistics and Inference: Estimation Methods*, Academic Press, Boston.

Balakrishnan, N. and Rao, C.R. (Eds.) (1998a) *Handbook of Statistics, Vol 16, Order Statistics: Theory and Methods*, Elsevier, Amsterdam.

Balakrishnan, N. and Rao, C.R. (Eds.) (1998b) *Handbook of Statistics, Vol 17, Order Statistics: Applications*, Elsevier, Amsterdam.

Balakrishnan, N. and Sultan, K.S. (1998) Recurrence relations and identities for moments of order statistics, in Balakrishnan, N. and Rao, C. R. (Eds.) *Handbook of Statistics, Vol 16, Order Statistics: Theory and Methods*, Elsevier, Amsterdam.

Balanda, K.P. and MacGillivray, H.L. (1988) Kurtosis: a critical view. *J. Amer. Statist. Assoc.*, 42, 111–119.

Bassett, G. and Koenker, R. (1982) An empirical quantile function for linear models with i.i.d. errors. *J. Amer. Statist. Assoc.*, **77**, 407–415.

Beirlant, J., Vynckier, P., and Teugels, J.L. (1996). Tail index estimation, pareto quantile plots and regression diagnostics. *J. Amer. Statist. Assoc.*, **91**, 1659–1667.

Boscovitch, R.J. (1757) De littreraria exedicione per pontificiamditionem, et synopsis amplioris operis ac habenturplura ejus ex examplaria etiam sensorum impressa Bononiensi Scientiarium et Artum Instituto Atque Academia Commentaii, **4**, 353–396.

Box, G.E.P. and Cox, D.R. (1964) An analysis of transformations. *J. Roy. Statist. Soc., B*, **26**, 211.

Box, G.E.P. and Muller, M.E. (1958) A note on the generation of random normal deviates. *Ann. Math. Statist.*, **29**, 610–611.

Brooker, J.M. and Ticknor, L.O. (1998) A brief overview of kurtosis, *Proc. Joint Statist. Mtg. Amer. Statist. Assoc.*, Dallas.

Burr, I.W. (1942) Cumulative frequency functions. *Ann. Math. Statist.*, **13**, 215–232.

Burr, I.W. (1968) On a general system of distributions III. The Sample range. *J. Amer. Statist. Assoc.*, **63**, 636–643.

Burr, I.W. (1973) Parameters for a general system of distributions to match a grid of α_3 and α_4, *Comm. Statist.*, **2**, 1–21.

Burr, I.W. and Cislak, P.J. (1968) On a general system of distributions I, its curve-shape characteristics. *J. Amer. Statist. Assoc.*, **63**, 627–635.

Bury, K.V. (1975) *Statistical Models in Applied Science*, John Wiley and Sons, New York.

Carmody, T.L., Eubank, R.L., and LaRiccia, V.N. (1984) A family of minimum quantile distance estimators for the three-parameter Weibull distribution, *Statistische Hefte*, **25**, 69–82.

Castillo, E. and Hadi, A.S. (1994) Parameter and quantile estimation for the generalized extreme-value distribution, *Econometrics*, **5**, 417–432.

Castillo, E. and Hadi, A.S. (1995) Modelling lifetime data with application to fatigue models. *J. Amer. Statist. Assoc.*, **90**, (431) 1041–1054.

Castillo, E. and Hadi, A.S. (1997) Fitting the generalized Pareto distribution to data, *J. Amer. Statist. Assoc.*, **92**, 1609–1620.

Castillo, E., Sarabia, J.M., and Hadi, A.S. (1997) Fitting continuous bivariate distributions to data, *The Statistician*, **46**, (3) 355–369.

Cheng, C. and Parzen, E. (1997) Unified estimators of smooth quantile and quantile density functions, *J. Statist. Planning Infer.*, **59**, (2) 291–307.

Chou, Y-M., Polansky, A.M., and Mason, R.L. (1998) Transforming non-normal data to normality in statistical process control, *J. Qual. Technol.*, **30** (2) 133–141.

Cox, D.R. and Hinkley, D.V. (1974) *Theoretical Statistics*, Chapman & Hall, London.

David, H.A. (1970) *Order Statistics*, John Wiley and Sons, New York.

Dielman, T, Lowry, C., and Pfaffenberger, R. (1994) A comparison of quantile estimators. *Commun. Statist. Simul, and Computation*, **23**, (2), 355–371.

Doumonceaux, R. and Antle, C.E. (1973) Discrimination between generalised extreme value, log normal and Weibull distributions, *Technometrics*, 15, 923–926.

Dudewicz, E.J. and Karian, Z.A. (1996) The extended generalized lambda distribution (EGLD) system for fitting distributions to data with moments, II: tables, *Amer. J. Math. Manage. Sci.*, **16**, (3 and 4), 271–332.

Dudewicz, E.J. and Karian, Z.A. (1999) Fitting the generalized lambda distribution(GLD) system by the method of percentiles, II: tables, *Amer. J. Math. Manage. Sci.*, **19**, 1–73

Dudewicz, E.J., Ramberg, J.S., and Tadikamalla, P.R. (1974) A distribution for data fitting and simulation, *Ann. Tech. Conf. Amer. Soc. Qual. Control*, **28**, 402–418.

Efron, B. and Tibshirani, R. (1993) *An Introduction to the Bootstrap*, Chapman & Hall, New York.

Evans, M., Hastings, N., and Peacock, B. (1993) *Statistical Distributions* (2nd ed.), John Wiley and Sons, New York.

Falk, M. (1997) On MAD and comedians, *Ann. Inst. Statist. Math.*, **45**, 615–644.

Falk, M. (1998) A note on the comedian for elliptical distributions, *J. Multivariate Anal.*, **67** (2) 306–317.

Fisher, R.A. and Tippet, L.H.C. (1928) Limiting forms of the frequency distribution of the largest or smallest member of a sample, *Proc. Cambridge Philos. Soc.*, **24**, 180–190.

Fowlkes, E.B. (1987) *A Folio of Distributions: A Collection of Theoretical Quantile-Quantile Plots*, Marcel Dekker, New York.

Freimer, M., Mudholkar, G.S., Kollia, G., and Lin, C.T. (1988) A study of the generalised Tukey lambda distribution, *Comm. Statist. Theory Math.*, **17**, (10), 3547–3567.

Galton, F. (1875) Statistics by intercomparison: with remarks on the Law of Frequency of Error, *Philos. Mag.*, 4th series, **49**, 33–46.

Galton, F. (1883) *Enquiries into Human Faculty and its Development*, Macmillan, London.

Galton, F. (1882) Report of the Anthropometric Committee, in *Rep. 51st Meet. Brit. Assoc. Advance. Sci.*, 1881, 245–260.

Gastworth, J.L. and Cohen, M. (1970) Small sample behaviour of some robust linear estimates of location, *J. Amer. Statist. Assoc.*, **65**, 946–973.

Gilchrist, W.G. (1976) *Statistical Forecasting*, John Wiley and Sons, Chichester.

Gilchrist, W.G. (1984) *Statistical Modelling*, John Wiley and Sons, Chichester.

Gilchrist, W.G. (1993) Modelling capability, *J. Opl. Res. Soc.*, **44**, (9) 909–923.

Gilchrist, W.G. (1997) Modelling with quantile distribution functions, *J. Appl. Statist.*, **24**, (1) 113–122.

Govindarajulu, Z. (1977) A class of distributions useful in life testing and reliability with applications to non-parametric testing, in Tsokos, C.P. and Shimi, I.N. (Eds.) *The Theory and Applications of Reliability*, Academic Press, New York.

Greenwood, J.A., Landwehr, J.M., Matalas, N.C., and Wallis, J.R. (1979) Probability weighted moments, *Water Resour. Res.*, **15**, (6) 1049–1054.

Grimshaw, S.D. and Ali, F.B. (1997) Control charts for quantile function values, *J. Qual. Tech.*, **29**, (1) 1–7.

Groeneveld, R.A. (1998) A class of quantile measures for kurtosis, *Amer. Statistician*, **51**, (4) 325–329.

Groeneveld, R.A. and Meeden, G. (1984) Measuring skewness and kurtosis, *The Amer. Statistician*, **33**, 391–399.

Hahn, G.J. and Shapiro, S.S. (1967) *Statistical Models in Engineering*, John Wiley and Sons, New York.

Hald, A. (1998) *A History of Mathematical Statistics from 1750 to 1930*, John Wiley and Sons, New York.

Harter, H.L. (1961) Expected values of normal order statistics, *Biometrika*, **48**, 151–157.

Harter, H.L. (1985) Method of least absolute values, in Kotz, S. and Johnson, N.L. (Eds.) *Encyclopedia of Statistical Sciences*, Vol 5, John Wiley and Sons, New York.

Harter, H.L. (1988) Weibull, log-Weibull and gamma order statistics, in Krishnaiah, P.R. and Rao, C.R. (Eds.) *Handbook of Statistics, Vol 7*, 433–466. Elsevier, Amsterdam.

Hastings, C., Mosteller, F., Tukey, J.W., and Winsor, C.P. (1947) Low moments for small samples: a comparative study of statistics, *Ann. Math. Statist.*, **18**, 413–426

Hazen, A. (1914) Storage to be provided in impounding reservoirs for municipal water supply, *Trans. Amer. Soc. Civil Eng.*, **77**, 1539–1640. Discussion, 1641–1649.

Healy, M.J.R. (1968) Multivariate normal plotting, *Appl. Statist.*, **17**, 157–161.

Hettmansperger, T.P. and Keenan, M.A. (1975) Tailweight, statistical inference and families of distributions–a brief survey, in G. P. Patil et al. (Eds.) *Statistical Distributions in Scientific Work, Vol 1*, D. Reidel Publishing, Dortrecht, Holland.

Hill, D.M. (1975) A simple approach to inference about the tail of a distribution, *Ann. Statist.*, **3**, 1163–1174.

Hogg, R.V. (1974) Adaptive robust procedures: a partial review and some suggestions for further applications and theory, *J. Amer. Statist. Assoc.*, **69**, 909–923.

Hosking, J.R.M. (1990) L-moments. analysis and estimation of distributions using linear combinations of order statistics, *J. Roy. Statist. Soc.*, **52**, (10), 105–124.

Hosking, J.R.M. (1998) L-estimation, in Balakrishnan, N. and Rao, C.R. (Eds.) *Handbook of Statistics, Vol 17*, Elsevier, Amsterdam.

Houghton, J.E. (1978) Birth of a parent: the Wakeby distribution for modelling flood flows, *Water Resour. Res.*, **14**, 1105–1110.

Huber, P.J. (1981) *Robust Statistics*, John Wiley and Sons, New York.

Hurley, C. and Modarras, R. (1995) Low storage quantile estimation. *Computational Statist.*, **10**, 311–325.

Hutson, A. (2000) A composite quantile function estimator with applications to bootstrapping, *J. Appl. Statist.*, **27**, (4), in press.

Hynderman, R.J. and Fan, Y.N. (1996) Sample quantiles in statistical packages, *Amer. Statist.*, **50**, (4) 361–365.

Jenkinson, A.F. (1955) A frequency distribution of the annual maximum (or minimum) values of meteorological elements, *Qtr. J. Roy. Met. Soc.*, **81**, 158–171.

Joanes, D.N. and Gill, C.A. (1998) Comparing measures of sample skewness and kurtosis, *The Statistician*, **47**, (1) 183–189.

Johnson, N.L. and Kotz, S. (1973) Extended and multivariate Tukey lambda, *Biometrika*, **60**, 655–661.

Johnson, N.L., Kotz, S., and Balakrishnan, N. (1994 and 1995). *Continuous Univariate Distributions* (2nd ed.), Vols 1 (1994) and 2 (1995), John Wiley and Sons, New York.

Joiner, B.L. and Rosenblatt, J.R. (1971) Some properties of the range in samples from Tukey's symmetric lambda distribution. *J. Amer. Statist. Assoc.*, **66**, 394–399.

Jones, M.C. and Daly, F. (1995) Density probability plots, *Commun. Statist: Simul. and Computation*, **24**, (4), 911–927.

Jung, S.H. (1996) Quasi-likelihood for median regression models, *J. Amer. Statist. Assoc.*, **91**, 251–257.

Kalbfleisch, J.G. (1985) *Probability and Statistical Inference, Vol 2, Statistical Inference*, 2nd ed. Springer-Verlag, New York.

Karian, Z.A. (1996) The extended generalised lambda distribution. *Communic., Statist., Simul. Comput.*, **25**, (3), 611–642.

Karian, Z.A., Dudewicz, E.J., and McDonald, P. (1996) The extended generalized lambda distribution (EGLD) system for fitting distributions to data: history, completion of theory, tables, applications, the final word on moment fits, *Commun. Statist. Simul. and Comput.*, **25**, (3), 611–642.

Karian, Z.A. and Dudewicz, E.J. (1999a) Fitting the generalised lambda distribution to data: a method based on percentiles, *Commun. Statist., Simul. and Comput.*, **28**, (3), 793–819.

Karian, Z.A. and Dudewicz, E.J. (1999b) Fitting the generalised lambda distribution to data: a method based on percentiles. II tables, *Amer. J. Math. and Manage. Sci.*, 19, 1–79.

Karian, Z.A. and Dudewicz, E.J. (2000) *Fitting Statistical Distributions to Data: The Generalised Lambda Distribution and the Generalised Bootstrap Methods*, CRC Press, Boca Raton, Florida.

Kendall, M.G. (1940) Note on the distribution of quantiles for large samples, *Suppl. J. Roy. Statist. Soc.*, 7, 83–85.

King, R.A.R. and MacGillivray, H.L. (1999) A starship estimation method for the generalised lambda distributions, *Austral. N. Zealand J. Statist.*, **41**, (3) 353–374.

Kodlin, D. (1967) A new response time distribution, *Biometrics*, 23, 227–239.

Koenker, R.W. and Bassett, G. (1978) Regression quantiles, *Econometrica*, **46**, (1) 33–50.

Koenker, R.W. and D'Orey, V. (1987) Computing regression quantiles, *Appl. Statist.*, **36**, 383–393.

Koenker, R.W. and D'Orey, V. (1993) Computing regression quantiles, *Appl. Statist.*, **43**, 410–414.

Kotz, S. and Johnson, N.L. (1985) *Encyclopaedia of Statistical Sciences*, John Wiley and Sons, New York.

Kotz, S. and Johnson, N.L. (1993) *Process Capability Indices*, Chapman & Hall, New York.

Landwehr, J.M. and Matalas, N.C. (1979) Probability weighted moments compared with some traditional techniques in estimating Gumbel parameters and quantiles, *Water Resour. Res.*, **15**, (5) 1055–1064.

Lanner, R. and Wilkinson, G. (1979) *Robustness in Statistics*, Academic Press, New York.

Laurie, D. and Hartley, H.O. (1972) Machine generation of order statistics for Monte Carlo computations, *Amer. Statist.*, 26, 26–27.

Lawless, J.E. (1982) *Statistical Models and Methods for Lifetime Data*, John Wiley and Sons, New York.

LeFante, J.L. and Inman, H.F. (1987) The generalised lambda distribution as a survival model, *Math. Biosci.*, 83,167–177.

Linhart, H. and Zucchini, W. (1986) *Model Selection*, John Wiley and Sons, New York.

Lloyd, E.H. (1952) Least squares estimation of location and scale parameters using order statistics, *Biometrika*, **39**, 88–95.

Ma, C. and Robinson, J. (1998) Approximations to distributions of sample quantiles, in Balakrishnan and Rao (Eds.) *Handbook of Statistics, Vol 16, Order Statistics Theory and Methods*, Elsevier, Amsterdam.

MacGillivray, H.L. (1992) Shape properties of the g- and h- and Johnson families, *Commun. Statist.–Theory and Methods*, **21**, 1233–1250.

McNichols, M. (1987) A comparison of the skewness of stock return distributions at earnings and non-earnings announcement dates, *J. Accounting Econ.*, **10**, 239–273.

Monti, K.L. (1995) Folded empirical distribution function curves, mountain plots, *Amer. Statist.*, 49, (4) 342–345.

Moors, J.J.A. (1988) A quantile alternative for kurtosis, *The Statistician*, **37**, 25–32.

Moors, J.J.A., Wagemakers, R.Th.A., Coenen, V.M.J., Heuts, R.M.J., and Janssens, M.J.B.T. (1996) Characterising systems of distributions by quantile measures, *Satistica Neerlandica*, **50**, (3), 417–430.

Mosteller, F., Fienberg, S.E., and Rourke, R.E.K. (1983) *Beginning Statistics with Data Analysis*, Addison-Wesley, Reading, Massachusetts.

Mudholkar, G.S., Kollia, G.D., Lin, C.T., and Patel, K.R. (1991) A graphical procedure for comparing goodness-of-fit tests, *J. Roy. Statist. Soc. B,* **53**, (1) 221–232.

Mudholkar, G.S., Srivastava, D.K., and Friemer, M. (1995) The exponentiated Weibull family: a reanalysis of the bus-motor-failure data, *Technometrics*, **37**, (4) 436–445.

Mudholkar, G.S. and Hutson, A.D. (1997) Improvements in the bias and precision of sample quartiles, *Statistics,* **30**, 239–257.

Nahmris, S. (1994) Demand estimation in lost sales inventory systems, *Naval Res. Logistics*, **41**, 739–757.

Okur, M.C. (1988) On fitting the generalized λ-distribution to air pollution data, *Atmospheric Environ.*, **22**, (11), 2569–2572.

Ozturk, A. and Dale, R.F. (1985) Least squares estimation of the parameters of the generalized lambda distribution, *Technometrics*, **27**, (1), 81–84.

Parzen, E. (1979) Nonparametric statistical data modelling, *J. Amer. Statist. Assoc.*, **74**, 105–121.

Parzen, E. (1993) Change P-P plot and continuous sample quantile functions, *Commun. Statist.*, **22**, 3287–3304.

Parzen, E. (1997) Concrete statistics, in Ghosh, S., Schucany, W.R., and Smith, W.B. (Eds.) *Statistics of Quality*, Marcel Dekker, New York.

Pearson, K. (1895) Skew variation in homogeneous material, *Philos. Trans. Roy. Soc. London, Ser. A*, **186**, 343–414.

Proctor, J.W. (1987) Estimation of two generalized curves using the Pearson system, *Proc. ASA Comput. Sec.*, 287–292.

Ramberg, J.S. and Schmeiser, B.W. (1974) An approximate method for generating asymmetric random variables, *Commun. ACM*, 17, (2), 78–82.

Ramberg, J.S. (1975) A probability distribution with application to Monte Carlo simulation studies, in Patil, G.P. et al. (Eds.) *Statistical Distributions in Scientific Work, Vol 2*, D. Reidel Publishing, Boston.

Ramberg, J.S., Tadikamalla, P.R., Dudewicz, E.J., and Mykytha, E.F. (1979) A probability distribution and its uses in fitting data, *Technometrics*, **21**, 201–214.

Reiss, R.D. (1989) *Approximate Distributions of Order Statistics*, Springer-Verlag, New York.

Renyi, A. (1953) On the theory of order statistics. *Acta. Math. Acad. Sci. Hungary*, 4, 191–231.

Robinson, M.E. and Tawn, J.A. (1995) Statistics for exceptional athletic records, *Appl. Statist.*, **44**, (4), 499–511.

Rousseeuw, P.J. (1984) Least median of squares regression, *J. Amer. Statist. Assoc.*, **79**, 871–880.

Rousseeuw, P.J. and Bassett, C.W. (1990) The remedian: a robust averaging method for large data sets, *J. Amer. Statist. Assoc.*, 85, 97–104.

Ruppert, D (1987) What is kurtosis? an influence function approach, *Amer. Statist.*, **41**, 1–5.

Sakear, S.K. and Wang, W. (1998) Estimation of scale parameter based on a fixed set of order statistics, in Balakrishnan, N. and Rao, C.R. (Eds) *Handbook of Statistics*, 17, Elsevier, Amsterdam.

Sarhan, A.E. and Greenberg, B.G. (1962) *Contributions to Order Statistics*, John Wiley and Sons, New York.

Scarf, P.A. (1991) Estimation for a four parameter generalised extreme value distribution, Technical Report MCS-91-17, Centre for OR and Applied Statistics, University of Salford, Salford, U.K.

Schucany, W.R. (1972) Order statistics in simulation, *J. Statist. Comp. Simul.*, 1, 281–286.

Shapiro, S. and Gross, A.J. (1981) *Statistical Modelling Techniques*, Marcel Dekker, New York.

Sheather, S.J. and Marron, J.S. (1990) Kernel quantile estimators, *J. Amer. Statist. Assoc.*, **85**, 410–416.

Sheppard, W.F. (1903) New tables of the probability integral, *Biometrika*, **2**, 174–190.

Stedinger, J.R. and Lu, L-H. (1995). Appraisal of regional and index flood quantile estimators, *Stochas. Hydraul. Hydrol.*, **9**, 49–75.

Srivastava, D.K., Mudholker, G.S., and Mudholkar, A. (1992) Assessing the significance of the difference between the quick estimates of location, *J. Appl. Statist.*, **19**, (3) 405–416.

Stigler, S.M. (1986) *The History of Statistics*, Harvard University Press, Cambridge, Massachusetts.

Stuart, A. and Ord, J.K. (1987) *Kendall's Advanced Theory of Statistics, Vol 1*, Charles Griffin, London.

Tawn, J. (1992) Estimating probabilities of extreme sea-levels, *Appl. Statist.*, **41**, (1), 77–93.

Teichroew, D. (1956) Tables of expected values of order statistics and products of order statistics for samples of size twenty or less from a normal distribution, *Ann. Math. Statist.*, **27**, 410–426.

Tukey, J.W. (1960) The practical relationship between the common transformations of percentages of counts and of amounts, Technical Report No 36, Statistical Techniques Research Group, Princeton University, Princeton, New Jersey.

Tukey, J. (1962) The future of data analysis, *Ann. Math. Statist.*, **33**, 1–67.

Tukey, J. (1970) *Exploratory Data Analysis*, (Vol 1), Limited preliminary edition, Addison Wesley, Reading, Massachusetts.

Velilla, S. (1993) Quantile-based estimation for the Box–Cox transformation in random samples, *Statist. Probab. Lett.*, **16**, 137–145.

Walpole, R.E. and Myers, R.H. (1989) *Probability and Statistics for Engineers and Scientists*, Macmillan, New York.

Weibull, W. (1939) Statistical theory of the strengths of materials, *Ingenioor Vetenskps Akadem. Handlingar*, **151**, 1–45.

Yang, S.S. (1985) A smooth non-parametric estimator of a quantile function, *J. Amer. Statist. Assoc.*, **80**, 1004–1011.

Ying, S., Jung, S.H., and Wei, L.J. (1995) Survival Analysis with Median Regression Models, *J. Amer. Statist. Assoc.*, **90**, 178–184.

Zrinji, Z. and Burn, D.H. (1996) Regional flood frequency with hierarchical region of influence, *J. Water. Res. Planning and Manage.*, **122**, (4), 245–252.

Index